*Buy Me!*
*Buy Me!*

ALSO BY JOANNE OPPENHEIM
WITH BARBARA BRENNER AND
BETTY BOEGEHOLD:

*Raising a Confident Child:*
*The Bank Street Year-by-Year Guide*

BABY
SHOPPER

# *Buy Me! Buy Me!*

## THE BANK STREET GUIDE TO CHOOSING TOYS FOR CHILDREN

*Joanne F. Oppenheim*

PANTHEON BOOKS
NEW YORK

 Published in the United States by Pantheon Books, a division of Random House, Inc., New York, and simultaneously in Canada by Random House of Canada Limited, Toronto.

For acknowledgment information please see the Directory at the back of this book.

LIBRARY OF CONGRESS CATALOGING-IN-PUBLICATION DATA
Oppenheim, Joanne F.
Buy Me! Buy Me!
Includes index.
1. Toys—Social aspects—United States.
2. Toy industry—Social aspects—United States.
3. Child development—United States.
4. Toys—United States—Purchasing.
I. Title.
HQ784.T68067 1987 688.7'2'029 87-43044
ISBN 0-394-75546-4

Manufactured in the United States of America

Designed by Beth Tondreau Design

First Edition

# *Contents*

*Acknowledgments* • *xv*

*Introduction: Why a Book about Toys?* • *xvii*

Toys—the Wonderful Tools of Childhood • *xix*
What This Book Is About • *xxi*
How the Toys Were Selected for *Buy Me! Buy Me!* • *xxiv*

*PART I: AN OVERVIEW OF TOYLAND* • *1*

*I. An Overview of Toyland* • *3*

The Great Toy Glut • *3*
The Toymaker's Role in the "Buy-Me" Syndrome • *6*

How Marketing Has Changed • *6*
Power Marketing • *7*
The New TV-Toy Connection • *8*
How Power Marketing Limits What You Buy • *10*
Supporting the Supports • *11*
Limiting What's Bought, Sold, and Made • *13*
Only the Giants Can Play • *13*
The Extension Plan • *15*
How the "Buy-Me" Syndrome Affects Children's Play • *16*
Open-Ended Playthings • *17*
From Transformation to Transformers • *17*
Collecting • *18*
Throw-Away Toys • *20*
Merchandising Feelings • *21*
Merchandising Fitness • *22*
Merchandising Learning • *24*
Merchandising Research • *26*
Merchandising Imagination • *27*
Redefining Age Appropriateness • *29*
Violence and Toys • *31*
New Twists That Escalate Violence • *34*
The Parent's Role in the "Buy-Me" Syndrome • *40*
Great Expectations • *40*
Gifts with a Guilt Edge: The Frequent-Flyer Syndrome • *41*
Highway Bribery • *42*
Nothing's Too Good for My Child • *43*
Handling the "Buy-Me" Syndrome • *44*
Setting Limits • *44*
Expanding Horizons • *45*
Providing Variety • *45*
Encouraging Inventiveness • *46*
All Licensed Toys Are Not Equal • *46*
All Toys Are Not Bought • *48*
All Gifts Are Not Equal; or, Gifts You'd Rather Do Without • *49*
Safety and False Assumptions • *49*
Accidents Related to Toys • *50*

The Parent's Role in Accident Prevention • *51*
Safety Checklist • *53*
Consumerism 101 • *53*
Learning to Choose • *55*

*PART TWO: TOYS BY AGES AND STAGES* • *59*

*II. Toys for Infants: The First Year* • *61*

Research and the Toymakers • *62*
Avoiding Too Much Too Soon • *63*
First Toys: Birth to Six Months • *63*
Mobiles • *64*
Mirrors • *66*
Lap Toys • *66*
Crib Gyms • *68*
Choosing Soft Toys for Infants; or, All Bears Are Not Equal • *69*
Have Toys, Will Travel • *72*
Six Months to Twelve Months • *73*
Manipulative Toys • *73*
Balls • *76*
Independence and Interdependence • *78*
Being Resourceful Is Playful • *80*
Playpens and Walkers • *80*
Bathtime • *81*
Looking Ahead • *82*

*III. Toys for Young Toddlers: The Second Year* • *83*

The Toddler's World • *83*
Setting the Stage for Learning • *84*
Siblings: Safety and Satisfaction • *86*
Toys That Match Toddler's New Abilities • *86*
Push before Pull • *88*

Balls • *90*
Climbing, Sliding, and Swinging • *90*
Dolls and Stuffed Animals • *92*
Handy Toys: Manipulatives • *94*
Hinged Toys • *95*
Shape Sorters • *96*
Blocks and Construction Toys • *97*
Tub and Water Toys • *100*
Books as Toys and More • *103*
Music, Singing, and Dancing • *103*
Early Pretend-Play • *104*
Early Art Exploration • *107*
Summing Up • *109*

*IV. Toys for Older Toddlers: Two to Three Years* • *111*

New Boundaries • *111*
New Use of Language • *112*
New Uses of Toys • *114*
Toys for Physical Development • *114*
Climbing, Balancing, and Swinging • *115*
Wheel Toys • *116*
Balls • *117*
Boxes and Other Interesting Spaces • *118*
Music and Movement • *119*
Sand and Water • *119*
Art Materials • *121*
Pretend-Play • *125*
Intrusions into Early Imaginative Play • *131*
Manipulatives and Puzzles • *132*
Blocks and Construction Toys • *135*
Books • *136*
Storage and Safety • *137*
The Parent's Role • *138*
Supporting New Social Skills • *140*

Making Choices • 142
How Smart Are Early Smarts? • 144
Summing Up • 145

*V. Toys for Preschoolers: Three to Five Years* • *146*

The Nature of Preschool Play • 146
The High Drama of Make-Believe • 147
Reality and Fantasy • 147
Where Do Fairy Tales and Fantasies Fit? • 148
Play and New Social Development • 150
Playing with Ideas and Words • 151
More Than ABC • 152
Unstructured Materials • 154
Symbolic Thinking • 154
The Push for Early Achievement • 155
The Social Dimension • 156
Personal Choices • 157
How Parents Are Handling Violence • 158
What We Know and Don't Know • 160
Does Banning Work? • 160
Kids and Commercials • 162
Toys for Threes, Fours, and Fives • 163
Dress-Up Clothes • 164
Props for Play • 164
Plush Animals • 170
Sex-Typed Toys • 171
Puppets • 172
Transportation Toys • 173
Blocks • 174
Props for Blocks • 177
Construction Toys • 179
Art Materials • 180
Finger Paints • 185
Clay and Modeling Materials • 186

Collages and Constructions • *187*
Music and Motion • *189*
Toys for Active Play, Indoors and Out • *191*
Pedal Toys • *192*
Gym Equipment • *195*
Balls • *197*
Water Play • *198*
Sandboxes and Sand Toys • *199*
Furniture for Play • *202*
Sit-Down Toys • *203*
Puzzles • *204*
Board Games • *206*
Manipulatives • *207*
Books, Tapes, and Talking Toys • *209*
Audio-Visuals for Preschoolers • *213*
Computers and Preschoolers • *213*
Summing Up • *216*

*VI. Toys for the Early School Years: Six to Seven Years* • *217*

The Need for Active Physical Play • *217*
Basic Equipment for Physical Development • *219*
Bikes • *219*
Other Wheeled Toys • *220*
Skates • *221*
Snow Gear • *223*
Balls and Ball Games • *223*
Fitness Equipment • *226*
Other Outdoor Equipment • *227*
Equipment for Outdoor Explorations • *227*
Pets • *231*
Providing Toys and Time for Solo Play • *232*
Pretend-Play • *232*
The Social Power of Toys • *234*

Contradictions • *235*
Dolls and Plush Animals • *236*
Mini-Worlds • *239*
Puppets • *240*
Construction Toys • *241*
Woodworking • *243*
Art Supplies • *244*
Crafts • *246*
Puzzles • *248*
Sit-Down Games • *250*
Audio Visual and Electronic Playthings • *252*
The Real Thing • *254*
Summing Up • *258*

*VII. Toys for the Middle Years: Eight to Eleven Years* • *259*

Collecting • *261*
Social Pressure • *263*
"But It's My Money!" • *265*
Setting Limits • *266*
Toys and Gender • *266*
Show Biz • *271*
Puppets • *272*
Transportation Toys • *272*
Sports Equipment • *273*
Crafts • *276*
Art Materials • *277*
Construction Toys • *279*
Cooking • *282*
Science Materials • *282*
Games • *284*
Puzzles and Manipulatives • *286*
Summing Up • *286*

Conclusion • *287*

*PART THREE: THE DIRECTORY* • *291*

Catalogues • *293*
Manufacturers and Distributors • *297*
Parent Action Groups • *301*

*Index* • *303*

# *Acknowledgments*

So many people have been helpful in the writing of this book. My thanks go to the parents and children of the Bank Street School for Children who responded to my lengthy questionnaires and to those who shared their experiences through interviews.

Thanks also to the many colleagues from the Bank Street College of Education Graduate School, Research Division, and Media Group who shared their insights and comments as the book took shape. I am particularly indebted to Dr. Leah Levinger, Dr. Edna Shapiro, Dr. Nancy Balaban, Elizabeth Gillkeson, Ellen Galinsky, James Levine, Barbara Brenner, and my editor, William H. Hooks.

I am grateful to Betsy Amster, who hammered out the structure of the book, to Jenna Laslocky, who held all the pieces together, and to Tom Engelhardt, who added his enthusiastic support and ideas.

Special thanks to my husband, Stephen L. Oppenheim, who lived with the stacks of catalogues, canceled weekends, and my preoccupation with the manuscript. He not only listened, he also asked hard questions that helped me to sharpen my thoughts.

Thanks also to the toy manufacturers and their public-relations representatives, who supplied photographs, and to Ann Darling Perryman and Mary Jo Oppenheim, who filled in the missing photos.

Grateful acknowledgment is made to Lorenzo Martinez, who typed the early drafts of the manuscript, and to Margaret Peet, who gave up weekends and evenings to meet the deadline.

# *Introduction*
# *Why a Book about Toys?*

To children there are few things so eagerly anticipated, fervently desired, or frequently requested as a new toy. Ask them what they like about toys and the answer is simple—toys are fun!

And they're right. Toys *are* fun. What's more, they are the basic tools children use to enrich their play and learning.

But today, with such a multitude of toys to choose from, finding your way through toyland can be both costly and confusing. From bitter experience you may know that far too many toys are less than wonderful, and that all too often the toys you buy are overpriced, easily broken, and quickly discarded. Given that, how do you go about choosing the best toys? And if you're a newcomer to toyland, how do you know what to select from the amazing array? If you're a veteran toy shopper, what do you do when you're torn between your child's plea for the best promoted toy and your own instincts about what would be the best, or at least a better, choice? This book can help you find your way.

At the same time it answers a host of other questions. Are some toys more basic than others? What can you do about TV and kids' appetites for everything they see? Do toys really teach, and if so, what lessons are built into the toys you select? Are there toys you should avoid? Can the violent content or themes of toys affect behavior? Is the quantity of toys you buy related to the quality of play? Can a child have too many toys? How can parents avoid or deal with the "buy-me's"? How do you find the right toys for a child's agegroup?

Much as we like to blame the toy companies and television for bombarding our children with messages to "Buy! Buy! Buy!" it may be that we as parents have played a part in fueling children's desires. Just visit a family with a new baby or young toddler and you'll see that long before children can form the words "Buy me!" the buying has begun.

In our eagerness to make our children happy, we've assumed that since toys make them happy, more toys will make them happier.

Because some toys can help children learn, we've assumed that more toys will make them smarter.

Because we want our children to have the best of everything we can provide, we've assumed that the best costs the most.

Because we want to show our children how much we love them even though we can't always be there, we've assumed that the more things we bring them, the better loved they'll feel.

Because we want to pat our children (and ourselves) on the back for doing well, acting considerately, and being brave, we've assumed that by rewarding such behavior with toys, the kids will be braver, kinder, and wiser.

Because we want our children to love us and we know

they love us when we bring new toys, we've assumed that more toys will make them love us more.

Yet all these assumptions seem to be flawed, particularly since the more toys we buy, the more we fuel the appetite for toys.

You've probably noticed that the most expensive toys don't hold children's interest any more than cheaper toys do. As far as they're concerned, the "best" toy is the new toy. Despite all our interest in educating children, the toys we consider educational often end up on the shelf.

You may also have noticed that by rewarding good behavior, the price of good behavior keeps going up. In time the pattern is reversed. A new toy must be promised to get behavior you once were able to expect. At the same time, children learn to equate love and attention with things rather than people.

Clearly, it appears that more and more buying leads to less and less satisfaction. Indeed, *buying* toys seems to be more interesting to many kids than actually playing with them!

It's not that children don't need toys. They do. But as parents we need a clearer understanding of the values of play and the kinds of playthings that support children's developmental needs.

## *TOYS—THE WONDERFUL TOOLS OF CHILDHOOD*

For children, playing with toys is not just something they do to fill up their time. It's through play that children learn about the world of people and things. Both the infant batting at the cradle gym and the schoolchild kicking a ball flex and stretch their muscles, gaining a sense of themselves as active and able-bodied doers. By passing a rattle from one hand to another or fitting the pieces of a puzzle or model together, they develop the fine motor skills that enable eyes and hands to work in concert.

But playing is more than a way of developing muscles and coordination. It's the frontier of socialization. Whether they are rolling a ball back and forth in a game of roly-poly or playing house, during play children learn the language of give and take, of communication, cooperation, and even negotiation. What begins as a duet between parent and child gradually grows into an orchestra that includes playmates, schoolmates, and teammates.

When they babble to their teddy bears or dramatize a space adventure, children are using play as a vehicle for traveling the boundaries of reality and fantasy. In pretend play they not only use their imagination and creative energies but also learn to understand themselves, their feelings, and others.

Watch the next time you see your child building with blocks or shaping a sandcastle. You'll see that it's through play that children translate the world to *child size* and manageable proportions, where they are in control. At the heart of play there are no fixed rules save those made by the player. It's through play that children exercise their problem-solving skills to solve intellectual puzzles—and also social, emotional, and physical ones. No single sphere is more important than another. Like the gears in a clock, they are not separate but interlocking pieces of the whole system.

By observing the multiple modes of playing, we can see that play means many things to children. As a result, there is no "right" set of toys for every child, but rather a need for variety. At any given moment, toys and play may engage a child's physical, intellectual, and emotional being. From their early years on, children need toys for quiet time and active time, toys for indoors and out, toys for playing alone and together.

Toys are much more than fun and the right toys become the wonderful tools of childhood.

## *WHAT THIS BOOK IS ABOUT*

*Buy Me! Buy Me!* is both a challenge to the hidden agendas of toyland and a reassuring guide through it.

The book is divided into three sections. "An Overview of Toyland," like an aerial photo, will give you a picture of the toy business today, showing you how an industry that racks up more than $12 billion in yearly sales is changing what children play with and even how they play.

The second section, "Toys by Ages and Stages," functions as a series of detailed roadmaps to the developmental milestones of childhood. Each chapter gives you the names of toys that are especially appropriate for the specific age and stage of your child's development and steers you clear of toys that are unnecessary or even undesirable. Since no book can list all the toys available, each chapter also gives you guidelines so that you can evaluate toy choices on your own:

- "Toys for Infants: The First Year" sorts out the toys you'll want for your child's early crib days through the sitting-up and standing-up stages. You'll discover when and how to stimulate your baby's rapidly changing social, physical, and intellectual development with a variety of playthings, including toys for the crib, changing table, stroller, high chair, carseat, tub, and lap games. At the same time we'll look at the ways toymakers have used and misused research to create unnecessary toys; and we'll talk about the difference between creating a stimulating environment and a frenzied one.
- "Toys for Young Toddlers: The Second Year" will help you zoom in on the special play needs of young

toddlers with their new mobility and appetite for exploration and independence. You'll find toys here that complement toddlers' growing repertoire of motor skills, together with new kinds of playthings for their early games of pretend. This chapter will tell you when and how to introduce a toddler to first explorations with art materials, music, and media. You'll also find ideas here for encouraging speech, curiosity, and imagination.

- In "Toys for Older Toddlers: Two to Three Years" you'll learn why toys for pretend play take on new importance. You'll discover how a toddler's expanding language skills enable new ways of thinking, learning, and playing. This chapter will help you choose playthings for solo play as well as toys for playing with others. You'll find toys that match your child's expanding social and intellectual skills as well as guidelines for choosing puzzles and manipulative toys that challenge problem-solving skills and eye-hand coordination. You'll also learn about the importance of materials for sensory and physical development.
- "Toys for Preschoolers: Three to Five Years" presents a greater number and variety of toys that reflect preschoolers' growing repertoire of play. You'll discover how to support pre-reading and pre-math skills with more than a set of magnetic letters and numbers, and learn why pretend play is important to children's learning about themselves and others. We'll consider what toymakers call "educational" and what lessons such toys teach and don't teach. You'll find practical ideas on how to cope with the age typical TV-toy-buy-me's. This chapter also includes a helpful discussion of He-Man and the fantasy themes that dominate TV-toys and the playlives of many children, together with information on violence and how it relates to toys and play.
- "Toys for the Early School Years: Six to Seven Years" explains the continuing value of play to your school-

children. You'll learn about the connections between physical play and the new social world the child wants to join. You'll discover games and toys that can help children refine writing, reading, and math skills in the context of play instead of endless drills. We'll examine further the role of violence and toys, and how your role is shifting. You'll learn strategies for dealing with TV and the "buy-me's." Finally, we'll discuss the limited themes of TV toys and propose ways you can fuel more imaginative play by buying less.

- In "Toys for the Middle Years: Eight to Eleven Years" we'll look at your child's final flings in toyland. You'll learn why kids at this stage often shift from one enthusiasm to another and how to cut your losses without limiting their opportunities to try new skills. You'll also learn why secret clubs, team sports, and the collector's bug are developmental signposts of the age and how they affect children's toy needs. We'll look at the connections between physical and social development and how you can support both group life and individual interests and talents. You'll learn how pretend-play changes, mixing adventure and fantasy with facts and reality. Finally, you'll find games for the family as well as toys and equipment for social and solo play and learning.

The third section of the book, The Directory, gives you the names and addresses of organizations and toy suppliers that will help you enhance your child's playlife.

In using the book you may want to go directly to the chapter related to your child's age group. Or you may prefer to read "An Overview of Toyland" (chapter 1) as background and then move on to the age chapter. Either way, *Buy Me! Buy Me!* will be a book to return to when birthdays, holidays, and toy-buying occasions arise.

## *HOW THE TOYS WERE SELECTED FOR* BUY ME! BUY ME!

Considering the thousands of toys available, the job of choosing a select few has been a challenging experience. Fortunately, I've had a lot of help. I've interviewed parents, teachers, psychologists, and child-development specialists from urban, suburban, and rural communities. I've spoken with toymakers, toy dealers, and the real experts—children.

The parents and students of the Bank Street School for Children offered a special kind of help by participating in a written survey that gave me detailed information on what they're buying, what they're not buying, and what they're buying but not using. The questionnaires—one for parents and another for children—provided me not only with a toy inventory but also with parents' views and concerns about toy-related issues.

In the final analysis, the toys shown here have been chosen to match children's needs and interests at various developmental stages, and all of them are well made and entertaining. My hope is that whether you're a new parent or an experienced (and somewhat disillusioned) toy shopper, *Buy Me! Buy Me!* will show you how to cut through the confusion and feel good about your choices.

ONE

# *An Overview of Toyland*

# I. *An Overview of Toyland*

## *THE GREAT TOY GLUT*

Anyone who lives with children is no stranger to toyland. You don't need to make a special trip to a toystore anymore. Toyland is with us wherever we go. Supermarkets, pharmacies, restaurants, museums, zoos, and even twenty-four-hour convenience stores have toys for sale. "Forget it," you may say firmly to your child. "I'm not buying any toys today," or "Put it back, you just got a toy. You don't need that!" Yet the toys on the shelves seem to trigger an endless round of "buy-me's" all the same.

Even if you never leave home, there's no way to escape the traps laid by toyland. Toys on TV cry out the same message of "Buy me! Buy me!" Coupons on cereal boxes and soup cans, in magazines and mail-order catalogues reinforce the "Buy me!" refrain. Visit a home with a new baby and you'll see that parents and grandparents are buying. They may be worried and confused about the values these toys are teaching, and feel manipulated by the blatant connection between the toy companies and commercial TV, but they are still buying! Never before in history have so many children had so many toys. Not only are there more toys to choose from, but parents are choosing to buy more.

You don't need a scientific study to prove that parents and children are caught up in the "buy-me!" syndrome. Just ask any middle-class child to tell you how many toys he's got in his room. Or better yet, watch a family checking out of a supermarket-style toystore with a loaded shopping cart. Shopping carts full of toys? Think about it. As a child, did you ever go to a toystore expecting to buy more than one toy at a time?

The world of toyland has changed radically over the past decade, affecting the way toys are bought, sold, and designed. These changes, in turn, are dominating children's playlives by shaping the options for toy choices. Although children today have more toys than former generations even dreamed of owning, too many of these toys actually limit rather than enrich the quality of their play.

Why? Well, despite the apparent glut of choices, the toys available are more alike than not. Multiple purchases have displaced the idea of multiple uses. The idea of using one plaything in many different ways is becoming more and more foreign. What child will devote himself to finding new uses for his old toys when he's constantly pressured to buy new TV-linked toys?

On a trip to toyland today, the greatest discovery is likely to be finding the exact location in the store of a toy touted on TV. Going to the toystore used to be an experience heightened by the possibility of surprise. As children we looked first in the window and then browsed in the store, and we were never sure what we'd find. But children today no longer ask simply for a doll or a plane. They want a Cabbage Patch Kid or a GI Joe Tomahawk helicopter. Preschoolers, many of whom no longer know Humpty Dumpty or Little Miss Muffet, can sing every word of the commercial jingles and theme songs of popular TV shows. The toy companies use TV as an electronic catalogue in your home.

While activist groups have brought attention to the link between toys, TV, and violence, the troubling issues in toyland are much broader than the question of banning or buying war toys. Indeed, the problems in toyland are not solely related to TV. It's not just children who are responding to the pressures of toyland. It's all of us. For parents the bombardment begins even as they decorate the nursery. Sign up for a prenatal class and you're on the list for newsletters and magazines, and dozens of companies who want to make your baby healthy, happy, and wise.

Long before babies start watching TV, the icons of Sesame Street and Disneyland are dangling on their crib rails and carriages. Before they are two, toddlers who don't yet speak in sentences point and say "Big Bird" and "Mickey Mouse." Their nursery toys are a preview of coming attractions in the world of licensed merchandise. You could call it the soft sell that comes before the hard sell. Quite apart from the question of good or bad, they are part of the reality of here and now, and part of the challenge of raising children in a consumer-driven world.

Marketing toys has become a highly sophisticated game. Before parents can negotiate the long aisles of today's toy supermarkets they need to know how the pressures of toyland affect both themselves and their children. They need to know how to sort out the mixed messages coming from the toymakers.

Selling toys is a lot like selling frozen diet dinners. Some ads sell the calorie count, others sell the flavor. Some try to do both. In the toy business, the pitch to parents is education, while the pitch to kids is entertainment. Some toys make claims for both.

Yet a closer look often reveals that what's pitched as educational frequently offers the most limited kinds of learning and play. Similarly, many supposedly entertaining toys hold limited interest and offer a strangely grim or saccharine notion of entertainment.

Of course, there are many fine toys out there that manage to do both. Good toys have always combined fun and discovery and lots of possibilities for imaginative play. Parents, then, have to look carefully beyond the advertising gimmicks that are running toyland today.

## *THE TOYMAKER'S ROLE IN THE "BUY-ME" SYNDROME*

### *How Marketing Has Changed*

In 1980 a doll named Strawberry Shortcake led a revolution that changed the way toys are created, marketed, bought, and even played with. The business of making merchandise from popular radio, TV, and movie characters is as old as Grandpa's Mickey Mouse watch or Grandma's Shirley Temple doll. But Strawberry Shortcake was no mere spin-off. Like Venus rising from the sea full grown, the little lady and her friends Lemon Meringue and Peaches 'n Cream arrived on the scene with a media splash that turned her into a household word. She starred in her own TV special, and her likeness was stamped on bedsheets, lunchboxes, pajamas, underwear, costumes, notebooks, greeting cards, party goods, picturebooks, and hundreds of other pieces of merchandise.

In the wake of Strawberry Shortcake, toymakers no longer think in terms of a new toy but rather of a new line.

This type of thinking defines the kinds of toys they produce, limits the kinds of toys we're buying, and ultimately constricts the quality of children's play itself.

## *Power Marketing*

Toymakers have always paid the owner of a character for the right to produce, say, a Howdy Doody puppet or a Superman costume. But Strawberry Shortcake was a whole new piece of cake. The American Greeting Card Company brought her to Kenner, and together the two companies orchestrated a new way of doing business. For the first time a character was invented specifically to become merchandise.

During her three years as a superstar, Strawberry Shortcake grossed a billion dollars. Seeing the strength of "power marketing," Mattel went to the drawing board and came up with He-Man, a small but mighty action figure who was to become not just the leader of the Masters of the Universe, but the muscleman of this new type of marketing.

Rather than paying for the use of someone else's character, big toy companies began with He-Man to create their own cast of characters with a story line that could be translated into TV specials and animated cartoon series with toys to match. With the promise of megabucks for promotion, other companies were eager to pay toymakers for the license to manufacture toothbrushes, clothing, books, bedding, and party goods. Hasbro's Transformers are a perfect example. In 1986, fifty-one companies were selling 121 products with the Transformer logo.

By 1986 the Toy Manufacturers of America estimated that close to 50 percent of the toy industry's sales were of licensed products. Indeed, according to three thousand retailers surveyed in *Toy and Hobby World,* the top ten toys on the Toy Hit Parade are all licensed characters with strong

*The toy industry's annual retail sales have grown from $7 billion in 1980 to more than $12 billion in 1985. Even allowing for inflation and changing demographics, the influence of the TV-toy connection is apparent in the growth of the industry.*

TV connections. These characters aren't simply advertised *between* cartoons. They are the featured players, the stars of their own specials and cartoon series.

## The New TV-Toy Connection

When you were growing up with Bulwinkle and Huckleberry Hound, the animation studios created characters and shows of their own and the major toy companies confined themselves to unrelated spot advertisements that fit between the cartoons. They were sponsors in the old sense of the word. Today, sponsorship has taken on new meaning. Toy companies themselves are sharing or bearing the cost of creating movies and cartoons featuring toys in their product lines. Nor does the cooperative venturing end there. The "you scratch my back, I'll scratch yours" approach extends to a third-party partner—the TV stations themselves. Through new barter arrangements, stations that carry syndicated cartoon series get not only the programs but in some cases a share of the profits from the toys they feature. This clever marketing device gives everybody involved a hand in everybody else's pocket. But, of course, the purse strings they're pulling on are on the other side of the TV screen—yours.

Given the snug relationship between TV and the toymakers, critics charge that most of today's shows are nothing more than thirty-minute commercials. "They sell products while claiming to be entertainment," claimed Dr. William H. Dietz, chairman of the American Academy of Pediatrics Task Force on Children and Television in a *New York Times* article. Dr. Dietz pointed out that if a show for adults were based on Hoover vacuum cleaners, it would be boycotted. In cautioning parents against so-called "product-driven" shows, he warned that they "hook kids on these program-length commercials that in fact offer an engaging story but are designed to sell one product . . . and kids don't know the difference. It is unfair and deceptive advertising. It is unethical to do that, in my opinion."

In its own defense the television industry cites research indicating that children do indeed know a legitimate program from an ad. They also point out that stations don't run ads for the same product that's being featured on a particular product-driven show. In other words, a Care Bears commercial doesn't appear on the Care Bears cartoon show. But most critics feel the line between programming and advertising is blurred. Young viewers may be able to tell where the so-called ads begin and end, but in fact it's all a commercial.

For television stations, film animators, and toymakers, this cozy arrangement is a gold mine in ratings and profits. Mattel's Masters of the Universe line raked in an estimated $350 million in worldwide sales in 1985. Hasbro's Transformers and Tonka's Gobots sold over $100 million worth of robots each in their first year out. All three products were supported with animated TV specials and syndicated series. With such returns is it any wonder that the TV-toy connection gets bigger and more incestuous each year?

Peggy Charren, president of Action for Children's Television, says that the commercial broadcasting industry "has given over children's television to the toy companies to produce."

Call product-driven programming exploitative, immoral, manipulative, whatever you like—it isn't illegal. But it used to be! In 1969 the FCC banned a program based on Mattel's Hot Wheels. They found that the show was "designed primarily to promote a product," not to entertain or inform. The current FCC, however, has thrown out the old guidelines. Deregulation is the order of the day.

This has opened the door to a new kind of TV-toy connection. Mattel's Captain Power and Axlon's Tech Force toys are actually activated by signals from cartoons on TV. It brings arcade-type games formerly played by much older

*Captain Power*
*Mattel*

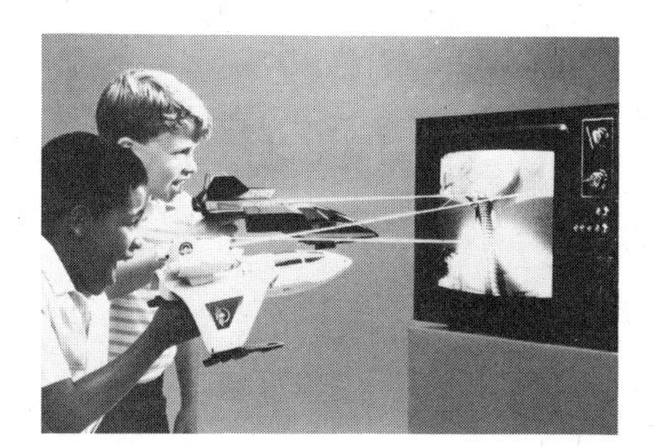

children directly into the home, making them accessible to a much younger audience. Though the technology may be innovative, the content continues to be based solely on violence. Despite the efforts of Action for Children's Television and other activist groups that have lobbied to halt the commercial exploitation of toy-based programming, the FCC has held that further government regulation is not needed.

It is their view that the marketplace dictates what constitutes the public's interest. In other words, if the public spends billions of dollars on TV-touted toys, the public is getting what it wants—the marketplace is speaking.

## *How Power Marketing Limits What You Buy*

Just as the megabuck marketing approach has constricted the kinds of programming found on children's television, the same mentality is creating a glut of toys that are more alike than not. The success of Masters of the Universe and its cohorts has spawned a legion of action figures cast in endless intergalactic battles that test the forces of good and evil. As one industry leader put it, "It's everyone against He-Man!" Transformers and Gobots added a new mechanical dimension. This year we have Visionaries and Supernaturals, action figures with holograms. But essentially the themes are more of the same. Like formulas for romance novels or whodunits, the names may change but the form and content are remarkably predictable.

Today, for a toymaker to be interested, toys must fit the stereotypical profile for an animated series: superheroes for boys and supersweets for girls. Originality or basic play value are less important than creating a cast of characters that can "communicate on TV." As Jeep Kuhn, an independent toy inventor, put it, "Today's toys begin with what looks good on television. To me that's a very negative thing. Instead of saying we ought to give them a good product, something that will be good for kids, or even something they'll enjoy—that isn't the big consideration. We have to give them something we can communicate to them in a certain way. A toy today has to fit what I call the television method—sixty-five episodes. They make stories and sell the characters in those stories. As a result, there's very little novelty in the toy itself. The novelty is in the story, a specified fantasy rather than the child's own."

By 1986 toy company executives at Tonka, Kenner, and Mattel openly acknowledged that they are in the "entertainment business." According to an article in *Children's Business,* Tonka has instructed employees to "think television." Terri Paulsen, Tonka's manager of promotional services, said, "Our marketing people can no longer *not* include [television] when they're thinking about a product."

**HE-MAN HAS THE POWER**

*Mattel sold 35 million of the action figures alone in 1984: 94,628 a day; 66.4 each minute.*
**Tom Engelhardt,**
**"The Shortcake Strategy"**

*If you lined up the 70 million 5½ inch Masters of the Universe figures sold since 1981 . . . they would extend from New York City to Los Angeles and back again.*
**Richard Weiner**
**of R. Weiner, Inc.,**
**promoters of He-Man**

## *Supporting the Supports*

Product support doesn't end with five-days-a-week programming, holiday specials, or full-length features at the movies. Like old-time screen stars, TV-toys have taken to the road, making personal appearances in major shopping malls from coast to coast.

Mattel's Mighty Toys Explorama brings larger-than-life play environments to malls, complete with live-action presentations, animation, and hands-on experience with new toys. In 1985, two and a half million children and their families got to meet such favorites as He-Man and Rainbow Brite. As a community-service tie-in, Mattel also sends a Children's Video ID Center. You bring a child and a tape and they'll record an ID that can be helpful in locating missing children. At the same time, Mattel gives away "wishbooks" full of rebate coupons that obviously pay off. Estimates are that toy stores and toy departments in malls experience a 70 percent increase in sales thanks to Explorama promotions.

But the hype doesn't stop there. Toy companies often hire pediatricians, child psychologists, and educational experts as "spokespeople" to go on media tours. Appearing on TV or radio talk shows, their messages are generally veiled but the objective is clear. Say you've got a new product called Bounce It, an attractive but not necessarily innovative bouncing ball. How do you promote it? It's simple. You hire a pediatrician who will appear on any number of talk shows to discuss the latest research on the importance of physical activity for young children's development. Your expert will also talk about the value of physical play for social and emotional growth. Somewhere during the five-minute spot that expert will also just happen to mention a perfect toy for social interaction, eye-hand coordination, and large-muscle action called Bounce It. Other toys may be mentioned in passing—but Bounce It will be named and probably shown.

Unlike a paid commercial endorsement, this type of commercial is dressed up as news or a public-service spot. It may even contain useful information. But it is nothing short of an ad in disguise. As viewers, we recognize the technique when we see an author, actor, or singer on camera. But when experts imparting child-care information appear, we don't always recognize or expect hidden messages.

And that's not all. Remember when toys were brought to school just for "show and tell"? No more. Recently Matchbox Toys, the major Robotech licensee, provided fourteen thousand schools with teaching materials that featured—you guessed it—Robotech characters from TV. According to their press release, close to a million first- and second-graders used these educational kits. Nor is Robotech's cast of characters the only one getting in through the back door of schools these days. Robotech is just one of many toys being "adapted" to teaching materials that will find their way into the classroom. They may be appealing to children, but should toy manufacturers be preparing curriculum materials when their intention is to make a profit?

## *Limiting What's Bought, Sold, and Made*

At the annual Toy Fair in New York City, where buyers and the press come to preview new lines, TV screens dominate the scene. To the retailer, the quality of the products themselves is hardly the whole picture. Just as important are the previews of movies, animated series, and commercials flickering on the screen. Buyers are told that this "vehicle" or that character will appear on upcoming shows—which means, of course, that kids are going to be looking for it!

Graphs posted on showroom walls and figures inside press kits let the buyers know how much money will be spent in magazines, newspapers, and on TV to presell the product line. This budget for media support determines what the buyers will buy to put on their shelves and ultimately what kids will be playing with.

## *Only the Giants Can Play*

While the possibility for profits has been compared to the GNP of many Third World countries, the sheer cost of launching a new product line tallies up to such stunning sums that only the biggest companies can compete.

According to the Television Bureau of Advertising, money spent on TV ads for toys is swiftly escalating. In 1985, toy and sporting goods manufacturers spent a total of $160,866,500 on commercial air time, an increase of 32 percent over the previous year. Hasbro, the biggest advertiser, increased its spending by 108 percent over the first nine months of 1984. Mattel spent $30,725,000; General Mills (Kenner/Parker) spent $16,015,600; Tonka's budget was a paltry $10,030,500. These figures are only for TV ads, not the shared cost of TV animation, newspaper and magazine ads, or product development.

Only the giants of the industry can afford these high stakes. Indeed, the giants have been swallowing up the second- and third-tier companies, and either reshaping old lines to conform or dropping them. Creative Playthings, a company that produced durable, high-quality toys, was bought up by CBS Toys. Before long, much of the Creative Playthings line either vanished or was redesigned with the "Hello Kitty" license stamped all over them. When Hello Kitty faded, many of the basic toys went with her.

In the past two years, Hasbro has gobbled up Milton Bradley, Playskool, and CBS Toys. But the giant toy companies can't afford to take on small, individual toy ideas, no matter how clever. Although inventors used to bring new toy "items" to the big companies, today it's nearly impossible, according to Larry Killgallon of Ohio Art Company. "Items" don't fit the needs for a product line; they don't fit the formula for a story with a cast of characters and lots of accessories. "It's a trap they've created," Killgallon says (in *Toy and Hobby World*). "Large companies must maintain large internal staffs to develop new product concepts."

Smaller companies can enter the fray, but they must be prepared to risk a wad on a single entry. Worlds of Wonder's Teddy Ruxpin, a high-tech talking bear from the Silicon Valley, caused a lot of chatter during Christmas of 1985. But the truth is, he didn't rake in $70 million in his first year by word of mouth alone. He had the full TV treatment, including ads and an animated special. In 1986 the Teddy Ruxpin line had new books, tapes, and add-on-plug-in plush "friends" that talk back to Teddy. With Teddy Ruxpin's great success, Worlds of Wonder executives believe they have created a new category—"electronic plush"

—rather than a two- or three-season "item." A category can go on for ten years, and in today's toyland that's an eternity.

More recently, the giant Toys-R-Us, in the retail end of the business, is playing a part in determining your toy choices. An order from Toys-R-Us is so crucial that some toy manufacturers show their prototypes to Toys-R-Us buyers in December. If a toy doesn't make it with them it won't even be shown at the Toy Fair in February. Walking the aisles of toy supermarkets, it's hard to imagine how they have become the ultimate arbiters and tastemakers. One may shop there for the discount but it's hard to miss the bad taste that prevails. Some of the finest toys are not even carried by these supermarkets.

## *The Extension Plan*

A large number of products introduced each year at the Toy Fair are really not new, but extensions to existing lines. Like turned-up cuffs on trousers, they give new life to a declining line, enabling it to stick around for another season (or two or three). Of course, each new season offers even more opportunities for new characters, vehicles, and accessories.

But even the life of a successful TV-toy is usually limited to three to five years. Five years from now no one under ten will know who Strawberry Shortcake was, though it's a fair bet that twenty years from now she'll have a nostalgic comeback. The Care Bears tried to stay on with a new feature film and the addition of "cousins" to the line. But this year they're gone, along with the Wuzzles, the once-mighty Gobots, Sectaurs, and so many others.

There are two forces at work here that fuel the need for novelty. On the one hand, toymakers believe that a new toy without TV support can't make it. On the other hand, toys that get their life from TV are almost certain to fade out in a few seasons.

Although Hasbro likes to claim that GI Joe is an exception and has been around as long as Mattel's Barbie, the original GI Joe of the 1960s took a dive after Vietnam. In fact, he went on an extended leave from 1978 to 1982. Those who knew him well know that the new guy is an imposter, not even half the size of the original eleven-and-a-half-inch hero with twenty-one moving parts and lifelike hair. And today's three-and-three-quarter-inch Real American Hero gets his stature from the bottom line: in 1984 he commanded $125 million in sales!

Barbie, unlike other toys of the television age, for more than a quarter of a century never had a feature film, an animated special, or a syndicated series. Mattel has had the good sense not to interfere with the many kinds of fantasy that several generations of girls have spun from their own imaginations. Whether you love her or loathe her, Barbie represents a measure of trust in a child's ability to create her own stories. Of course, Barbie's "environments"—her sauna, office, kitchen, sportscar, and wardrobe—are part of her own extension plan. They provide a setting that shapes her story line without a show. Her props are pitched on TV spots that keep Barbie in style and on camera. Her longevity may very well be explained by the fact that without a TV series she leaves something for the imagination. However, current plans do include a live-action, magazine-style TV show with a "real-life" Barbie in the anchor seat. It will be interesting to see if Barbie's show affects her longevity.

## *HOW THE "BUY-ME" SYNDROME AFFECTS CHILDREN'S PLAY*

In examining the nature of the toy business we have seen how power marketing and TV have changed what's made and how it's sold. But the impact of those changes goes beyond profits and losses. They have a powerful effect on the toys children play with, and even how they play!

## *Open-Ended Playthings*

For years specialists have been advising parents to look for "open-ended" toys—in other words, toys that can be used again and again in multiple ways. A set of blocks may become a skyscraper today, a zoo tomorrow, and a space station the next day. Compared to a plastic castle that always looks like a castle, the unstructured nature of blocks allows children to shape as many new settings as their imaginations can invent. Likewise, a blank pad of coloring paper and crayons are more open-ended than a coloring book with pre-made pictures.

Today's toymakers, though, have redefined the meaning of open-ended to mean that *you're never finished buying.* What can you do with just one GI Joe? To play you need a battalion, if not a division. Similarly, you don't need just one plane, you need a fleet of planes, and a convoy of trucks, jeeps, and tanks, and maybe even an aircraft carrier.

It's the same with Barbie, He-Man, and Transformers. How can you play good guys versus bad guys if you don't have a large enough cast? When kids say they *need* it, they are not far from wrong.

## *From Transformation to Transformers*

Borrowing from the academic research on play, the toymakers have reshaped the notion of transformations. Anyone who has watched a child pick up a pencil and buzz it through the air like a plane is witnessing a true transformation. The ability to make one thing stand for another is no small intellectual feat. In making such transformations children begin to "play" with symbols—an ability that will eventually be needed to make symbols like *c-a-t* stand for the four-legged furry creature we call cat. Very young children need realistic toys for their symbolic play: for a toddler a toy phone must look like a real phone. The preschooler, on the other hand, can pick up a phone out of thin air and carry on a conversation.

Compared to the multiple transformations children are capable of making, the new transformable toys are more mechanical than imaginative. The ready-made Gobot or Transformer is either a robot or a vehicle, period. While such toys may challenge eye-hand coordination and manipulative skills, their precast identities represent a narrowing of play possibilities. Children who grow up with a trained dependence on literal play props may not move on to the more advanced level of playful flights of imagination.

After the success of Transformers and Gobots, the toy companies have rushed to translate the transformation idea into other categories. Now we have Popples, plush balls that turn into dolls; Jem, a punk-rock singer who turns into a funkie record executive; and plastic rocks that open to reveal action figures. While the "magic" may have novelty appeal, the transformation is no longer made by the child, but by the toymaker. A few of these toys will probably do no damage to children's play, but they should not predominate.

## Collecting

The collector's bug, the passion to amass baseball cards, bottle caps, stones, or political buttons, was until recently a hallmark of the schoolage child. This often short-lived but passionate interest was usually satisfied by swapping relatively inexpensive items or found objects. These days the appetite for collecting begins earlier and can cost a great deal. Whereas old collections were by-products—the lid of an ice cream cup, a stamp from a letter, a souvenir menu—the new collectibles are designed to be *buy*-products. Even at a discount, the cost of a stable full of My Little Ponies ain't hay. And you need a lot of spending power to assemble the full cast of Masters of the Universe.

Recognizing children's interest in collecting, Mattel recently introduced a barrelful of plastic miniature figures for children to buy and trade. A clever poster picturing all two

*M.U.S.C.L.E.s*
*Mattel*

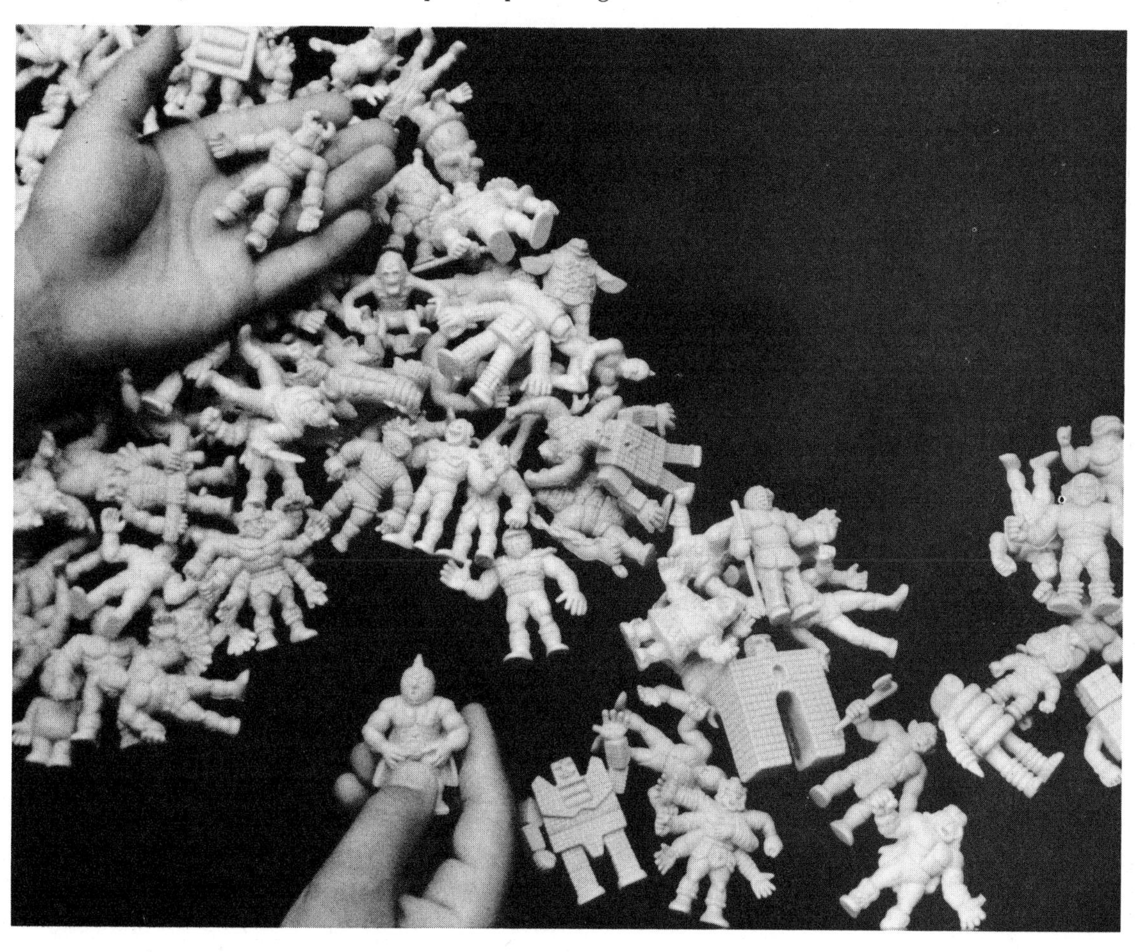

hundred and thirty-three characters is given away at the point of purchase so that collectors of M.U.S.C.L.E.s can check the ones they own and see at a glance which they still need. At twenty-five cents apiece the total collection costs $58.25, but compared to action figures, M.U.S.C.L.E.s are a bargain. However, even if money is no object, should childhood collecting be prepackaged? Isn't there something qualitatively different here?

Collections of stamps, baseball cards, coins, and postcards often lead to broader and sometimes lasting interests in sports, history, places, and people. They are part of the middle years' appetite for knowing and owning a piece of the bigger world. In contrast, prepackaged collectibles hardly lead to anything more than buying more of the same.

While manufactured collectibles may have some socially redeeming features as a tradable item in the schoolyard, for preschoolers they are another matter. The price of each item may be low, but in the long run they may give preschoolers a headstart in early consumerism.

## *Throw-Away Toys*

TV itself has a tremendous need for more and more shows. Characters get "used up" on both adult TV and children's programming. The media's need for new shows feeds the toy industry's annual parade of new products and ultimately both TV and toymakers feed kids' appetites for new "buy-me's."

In 1982, when I wrote a toy roundup for *Working Mother* magazine, I worked with press kits and catalogues. That article listed basic toys that had stood the test of time and kept on playing. The following year I covered the Toy Fair in person and was exposed for the first time to the glitz and blitz of public-relations reps pushing their new lines. That was the year Cabbage Patch Kids were born. "Put her on the cover of your magazine!" a Coleco rep urged me and every other reporter. *Working Mother* didn't but *Newsweek* did.

By the next year even I was walking in and asking, "What's new?" But is new better? Maybe not, but what's old doesn't stay around for very long anymore. Forgotten completely is the fact that to a six-year-old a toy staple like Lincoln Logs is a new and challenging building set. Yet each year a few more old favorites disappear. Even more interesting is the fact that many of the hottest items from the 1983 press kits have fizzled out of sight or are fading fast.

Children today rarely live with a toy long enough to invest themselves in it. By four or five, kids are learning to become junior consumers of a "throw-away world." When toys are bought by the bagful only to be bundled up and chucked out to make room for a new batch, is it any surprise that children's interest in buying overshadows their interest in play? For too many children the best toy is a new toy—the one they're going to buy.

## *Merchandising Feelings*

Taking their inspiration from the greeting card mentality of manufactured sentiments, the toymakers have attempted to shift the emotional content of play from the child to the toy. No longer are children expected to invest their own feelings in play. Instead, "feelings" are packaged for them—as merchandise.

Consider the Care Bears. Except for their color and the patches on their tummies, they all look very much the same. Unlike Snow White's Seven Dwarfs, each of whom had separate and easily distinguishable looks and personalities, the only real difference between the Bears is what's written on their labels. Kenner's Fluppy Puppies and their Woof Wear are also cast from the same mold. Looking at them it's hard to know just who's who. Put Fanci Flup, Shy Flup, Cool

Flup, Brave Flup, Cuddle Flup, and Silly Flup in a black-and-white glossy and you get the picture. Never mind, the Disney Fluppy Puppies cartoon will no doubt spell it all out.

Unfortunately, packaging feelings like this pulls the rug right out from under children's play. No doubt you've seen children bake a mud pie and smash it, or transform themselves into roaring lions or babbling babies. You've also probably seem them step into Mommy's high heels or blast off into space with the courage of an astronaut. It is in spontaneous, imaginative play that children try out all sorts of feelings, in the company of other children or in solitary play with an imaginary playmate or a teddy bear or doll. A teddy bear or monkey may be treated as if he's naughty, happy, silly, quiet, lovable, angry, shy, noisy, or whatever else the child wants or needs it to be at the moment. *The point is, the toy doesn't have feelings. The child does!*

Children don't learn about feelings from a toy's name or label. They learn about their own feelings and others from the people in their lives and the way they treat each other. A toy may come to represent comfort and security. In times of need it will be a nonjudgmental listener, a confidant, and a dependable friend. But those are qualities the child conveys, not the package. And it is the child alone who has the power to endow the toy with such qualities.

*AG Bear*
*Axlon*

## *Merchandising Fitness*

One of the most significant values of play centers on physical development. The day a baby first rolls over, sits up, stands, or walks is a momentous milestone. In the progression from physical dependence to increasing independence, the child gropes for ever-expanding skills. Indeed, it's in the backyard and at the playground that children continue to stretch and test their "can-do" sense of themselves. While parents may value intellectual achievement, the first frontier in the lives of children is more closely related to their physical competence.

Yet oddly enough, two of the most basic toys for active play are fast becoming the Edsels of toyland. Toddler-size riding toys driven by foot power and the classic preschool trike run on pedal power are both being replaced by battery-operated vehicles that exercise only the finger that pushes the stop/go button.

According to one manufacturer, pedal toys are "obsolete." Their research shows that "consumers," as they put it, prefer the fantasy play value of motorized cycles and cars. As a result, another dimension of vigorous, active play is being replaced by a more passive mode. For twice as much money you can buy less than half as much play. Seated in their red Corniche convertibles, children seem well on the way to yuppiedom. As soon as they're old enough, they can ride a stationary bike or join a health club.

In fact, formalized fitness programs for babies are another new trend in toyland. Plugging into gym classes that have become so popular for preschoolers, the toy companies are producing exercise mats and videos to use with baby. Again, the toymakers are taking an idea for older kids and promoting it for younger kids.

Although the programs (on tape or in classes) call for physical action and some interaction, the need for such programs may have more to do with trendiness than with children's actual needs. Given the isolation of suburban and apartment life, such classes may satisfy the need for social contact more than the need for exercise. Unlike the spontaneous play children engage in around the house and neighborhood, these exercise classes are simply another kind of programmed entertainment, with adults leading parents and children through their paces.

While adults who sit at desks all day may need a "workout" to overcome their passivity, young children shouldn't need a scheduled time for running, jumping, and flexing their muscles. A class once a week or a video tape is not an inoculation against passivity. Active play should be a part of their everyday experience.

## *Merchandising Learning*

Always on the lookout for a new marketing angle, the toy-makers have taken bits and pieces of information from the academic literature on play and attempted to turn them into merchandise and slogans. Look at the thrust of these ads: "Get SMART and watch how fast they get smart too!" . . . "If you take their fun seriously" . . . "Toys that allow your baby to master new skills" . . . "The first learning toy that plays like a friend" . . . "The very best fun your baby could ever learn from." It's not the toys we're looking at here, it's the message. The message that "play is the work of childhood" has been popularized into a cliché. In the name of play and learning, the work ethic has been brought into the nursery. Play is no longer valued for its own sake, but rather as a means to an end. If play is children's work, then children are expected to work at their play. Their playthings are expected to teach, and the sooner the better!

Plugging into the current craze for early smarts are baby bumpers emblazoned with numerals, color words, and geometric shapes. Indeed, to look at the glut of infant and toddler shape-sorting toys, one must wonder how and why such abstract forms fit the active infant's interest or needs. It's not that toddlers don't enjoy the fill-and-dump or twist-and-turn action of shape-sorters, but how many do they need? To the behavioral psychologist in a lab these may be useful tools for measuring cognitive development. But is there anything gained by turning test materials into play equipment? Does learning the names of shapes and letters early produce better babies and smarter children? There's no hard evidence to support that idea; in fact, some research indicates that this kind of abstract learning has more meaning later.

Playing on parental anxiety and eagerness to teach, the toy companies have come up with educational toys that are anything but smart teaching tools. Many of the newest electronic toys operate on the premise that learning can be reduced to right and wrong answers. Such a limited view of learning fits well with the notion that the most urgent curriculum of childhood can be reduced to knowing their ABCs and one-two-threes. However, the rush to teach formal lessons in reading, counting, and abstract symbols to babies, toddlers, and preschoolers fits neither their learning style nor other more immediate needs.

Is there any reason to put a toy clock on an infant's crib rail? Do two-year-olds need to learn how to tell time? Do three-year-olds need to start reading?

Playskool's *Teach Me Reader* is a perfect example of putting technology before content. Touted as a state-of-the-art reading development tool for preschoolers (ages three to six), the built-in curriculum is about as age-appropriate for them as the *Encyclopaedia Britannica.* Based on the sight method of teaching reading, the 450-word vocabulary of the *Teach Me Reader* would be a sore trial for many a second-grader. But quite aside from the overload, there's the false assumption here that three-, four-, and five-year-olds are ready to learn the formal mechanics of reading. Given that most educators today advocate a combined phonics and sight program for beginning reading, this totally see-say-hear method is hardly state-of-the-art. What a child gets is a flat, expressionless, disjointed word-by-word rendition of the story, a poor model for beginning readers. Nor are there many responsible educators who believe that learning to read 450 words is a useful skill for children who are only three, four, five, or six years old. Even as a tool for older children, say first- and second-graders, the program has serious limitations. A child who has mastered hundreds of sight words shouldn't be reduced to reading or hearing one word at a time, but rather helped to sweep a line of text and read in phrases.

Of course, electronic toys are not all equal, and many of them have great appeal to children. For drill and practice, such toys put the child in control of the pace and save him from losing face with parents and peers when he makes mistakes. But electronic teaching machines can't answer questions or interpret why a mistake is made or repeated.

During the preschool and early school years when children are learning to learn, parents need to guard against heavy emphasis on symbols, quick answers, and rote learning. Knowing how to count to one hundred is of little value if you don't know the meaning of the numbers. Nor should the importance of learning to question as well as answer be overlooked.

In rushing to teach abstract skills to preschoolers, the price of the effort may be higher than its worth in the long run. Sure, young children can learn the names of shapes and numbers and even words. But they are likely to learn the same things much faster and with greater ease a few years later. Long before children are ready for the rigors of learning to read, parents should be building a love of stories and books. Indeed, if you really are concerned about what makes good readers, there is nothing that measures up to the experience of good books.

## *Merchandising Research*

Now even research testing tools are being dressed up as toys.

Consider Johnson and Johnson's voice-activated mobile. Is something valuable happening when an infant's vocalization sets off a lullaby? For the researcher the study of infants' responses to sound, touch, and sight stimuli may be a good way of measuring and comparing data. But translating those techniques into playthings doesn't always lead to playthings that enhance development. Indeed, the artificial stimuli of the laboratory may distract from the real sights and sounds and responses an infant needs to learn. A baby's early sounds are by nature designed to communicate his needs to others—not to things, but to people. In addition to sending off messages of hunger and discomfort, his vocalizations are a way of exploring sound itself, and of trying out the way sounds are made and repeated. These are pieces of the process of acquiring speech. Now, one must wonder if something is gained by covering these sounds with a distracting lullaby. Do infants really need toys that intrude on the ways they play with their own voices?

While the same toy might delight and empower a toddler with a wonderful sense of making things happen, for the infant it may be of questionable value—not to mention a nuisance whenever an adult speaks. Although many of the Johnson and Johnson toys fit right into the infant's and toddler's emerging developmental skills, the mobile has an air of the behaviorist's lab about it.

Increasingly toymakers have promoted toys which, like the mobile, are labeled "interactive." But interaction in play used to refer to the child and parent or the child and other children. Today interactive is more likely to refer to the child and an automated toy. The toy has become the other player.

## *Merchandising Imagination*

Perhaps no word is more overused in toy advertising than *imagination.* Yet in packaging this important aspect of play, the child's role is too often diminished. Taken out of context, even a good concept can be translated into a poor toy. For example, miniature environments such as farms, fire houses, and garages have traditionally offered children realistic props for imaginative dramatic play. My own children had a sturdy Noah's ark with a menagerie of animals that were also used to play circus, zoo, and jungle adventures. The pieces were not locked in. Often such toys are used along with building blocks as part of a larger setting. Creating, directing, and playing out their little dramas, children use such pieces in flexible ways.

Consider, then, what happens when a mini-environment like Shuffletown comes packaged with all the pieces locked into a track. Sure, it's neater and there are no lost pieces. That you can turn it upside down as they do on TV is a selling point to mothers, but what's lost is the toy's basic play value. The Shuffletown vehicles and people can be manipulated only within the confines of the preset path. For a bedridden or physically handicapped child such toys may be useful. For the great majority of children, however, this kind of toy diminishes the possibilities of play. Indeed, it distorts a basically sound play concept.

*Power Shirt*
*Dynatec International*

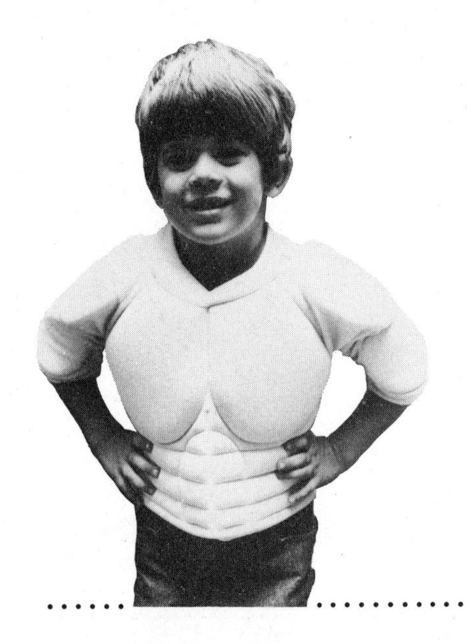

Still more troubling are the vast number of toys tied into TV cartoons. Are the elaborate stories, settings, and characters changing the quality of children's play? When manufacturers give action figures names and spell out their personalities on the back of their shrink-wrapped packages, are they also shrinking the child's imagination and play? Many educators think that it does, but we don't really know. This is an area that needs more research.

We do know that toddlers and preschoolers need highly realistic props for pretend play, but as they grow older the need for such props should diminish. Yet, today the dominant toys sold for young and old alike are more literal than ever. An action figure that's cast as a bad guy has an identity and role to play that's hard to override. These days they're selling padded Power Shirts so boys of five or six can don artificial pecs to become larger than life-size heroes. Next maybe they'll sell a Dolly Parton padded bra for five-year-old girls. It's as if we don't trust children today to spin their own fantasies and make their own transformations. Children who are dependent on this kind of elaborate tooling for play may never get beyond the level of replaying what toymakers have imagined for them.

Similarly, too much reliance on phones that talk, dolls that crawl, and toys that do the playing end up casting children in the role of watcher—of reactor rather than active doer. For some children such toys build an expectation of "what can the toy do?" rather than "what can I do with this toy?" Toys with limited agendas quickly lose their appeal. As a result, the more toys children have that work only one way, the more toys they're likely to need. It's no wonder that you're constantly treated to choruses of "buy me this" and "buy me that."

## *Redefining Age Appropriateness*

Increasingly, toys that might be relatively acceptable for older children are finding their way into the hands of the very young. "Age appropriate" has come to mean one-size-fits-all.

Transformers and robots have been so popular with schoolage kids that the toy companies went all out and produced oversized and simplified Transformers for toddlers. Yet the meaning a robot has for a two-year-old is hard to understand.

In general, age labels are of little use. The toymakers stretch the numbers on the box as far up and down as they can go. That's good marketing. But quite apart from safety considerations, which we'll discuss later, let's consider the content. Fashion dolls or action figures that are hard for little fingers to manipulate may nevertheless be labeled four and up. Never mind that the fantasies of Barbie or GI Joe are beyond the grasp of a preschooler—if the toy can't choke them, it's passed the test.

Unfortunately, television programming in particular encourages this one-size-fits-all mentality, which is reflected in the toys kids from two to twelve are playing with these days. Although the stations schedule programs for the youngest children early on Saturday morning, there is no way (short of constant parental involvement) of controlling who is watching what. As a result, age appropriateness no longer means much when it comes to stories, themes, characters, and, ultimately, toys.

TV animators (and toymakers) like to say that their products are no different than fairy tales. Forgotten completely is the fact that the themes and content of traditional fairy tales are anything but appropriate for children as young as three, four, or even five.

For preschoolers, who have a limited grasp of what's real and what's not, the "in" programs lead to "in" toys that may also be "in"-appropriate. The fact that a child watches intensely doesn't make a program or its related toys appropriate. Young children are often fascinated by that which puzzles or even frightens them the most. For some children, playing with a grotesque two-headed monster may represent a way of taming anxiety. For many others it may be a compelling fascination that induces anxiety.

Children's individual responses and needs can't be dismissed. For some children, action adventure programming and toys are overstimulating. Studies indicate that shows that play on a sense of great danger and unexplained monsters or supernatural characters may provoke nightmares or anxiety in some children. In order to determine the best approach to take with their own children, parents need to pay careful attention to their children's reactions.

Unfortunately, the overpowering appeal of power marketing is limiting the variety of play materials children have available to them. Toys that relate to playing house, doctor, fire fighting, and road building—in other words, reality-based pretend-play—are no longer the predominant toys purchased for preschoolers. Despite the fact that these are the basic play materials that help children understand the real world, they have been displaced by far-flung fantasy worlds that involve limited and essentially violent themes.

## *Violence and Toys*

To judge from the ratings, it seems fair to say that nothing is more appealing, seductive, or entertaining than violence. Five of the six hottest shows and toys in 1986 revolved around violence: Transformers, Masters of the Universe, Gobots, Voltron, and GI Joe. The toys were so popular that manufacturers couldn't keep up with the demand. A commercial aired in prime time before Thanksgiving warned parents to get to the toystore early for Transformers to avoid holiday disappointment.

In just two years, the sale of war toys has soared by 350 percent, and the quantity and level of violence is escalating rapidly. Mattel's He-Man looks almost mild compared to the new breed desperately trying to overtake him. The new levels of tastelessness are almost an attack on children's sensibilities. Just to give you a sense of what is being offered, consider the newest casts of characters introduced recently, shown on the next pages.

One can imagine the creative teams laughing at their own cleverness in coming up with such colorful names as Dr. Hacker, Bonecrusher, and D. Compose. But do plays on words, like Apokolips, lead to sound play material? Isn't this adult humor a bit cynical and meanspirited? Isn't it misplaced as children's entertainment?

This inventory of "heroes" points up what a limited menu children are being served. Are these heroes and heroines cast again and again because they satisfy? Or because they sell? Would more variety sell if it were offered?

***Dr. Terror, a Centurion of the 21st century, who has escaped with the top-secret Military Programs that will fuse Man and Machine. (Kenner)***

*Dr. Terror*
*Kenner*

***D. Compose, of the Inhumanoids, creatures from beneath the earth who want to control the world and destroy mankind. His foul stench can shrivel plants. His touch turns anyone into a rotting skeletal mutation. His hinged rib cage imprisons his victims. (Hasbro)***

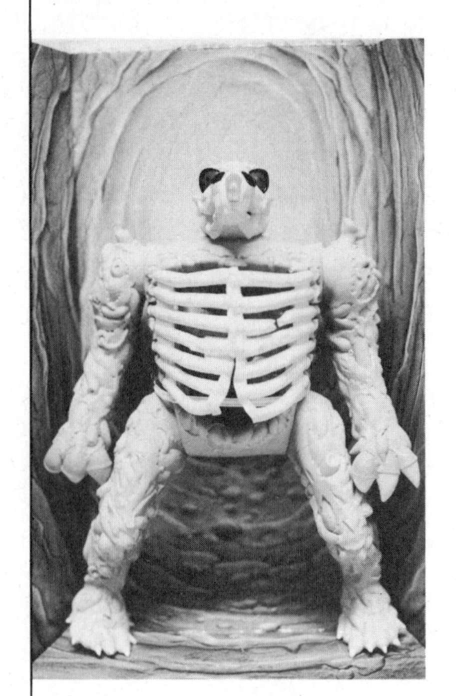

*D. Compose*
*Hasbro*

***Mumm-Ra, evil mutant and enemy of the Thundercats, with Battle-matic Action, whose quest is to capture the power of the Sword of Omens. Labeled ages 4 and up. (LJN)***

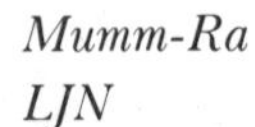

*Mumm-Ra*
*LJN*

***Sand Storm, hideous henchman of Tex Hex, enemy of Marshal BraveStarr, the toughest, bravest space lawman of the spaceport town, Fort Kerium. Fill the bold and bad Sand Storm with water and he'll blast the enemy with knockout vapors. (Mattel)***

*Sand Storm*
*Mattel*

*Buzz Saw*
*Hordak*
*Mattel*

***Buzz Saw Hordak, enemy of He-Man. His secret saw blades burst out of his chest to blast good guys like Snout Spout, Rio Blast, or Ram-Man. (Mattel)***

***Evil-Lyn, warrior goddess of the Evil Horde. Twist her waist and she packs a punch. But unlike male action figures she has no sword—only a crystal ball that helps her foresee the evil future. (Mattel)***

*Evil-Lyn*
*Mattel*

## *New Twists That Escalate Violence*

Increasingly, the toymakers and animators are repackaging entertainment designed for adults as merchandise for kids. Among the "heroes" who came directly from movies made for adult audiences to do battle with GI Joe are LJN's Delta Force and Coleco's Rambo. In fact, Coleco's attempts to cleanse Rambo of his bloody reputation brings to mind the handwringing Lady Macbeth. As a toy, Rambo is being billed as a true patriot, "the kind of person we can all turn to in times of trouble." As the press release says, "Our beloved country has been thrown into peril by international terrorists, commanded by the evil General Warhawk and his team, the S.A.V.A.G.E. terrorists." Rambo to the rescue.

Each year the quality of violence seems to escalate. Violence has changed from old-fashioned war games and shoot-'em-ups to assaults on the child's emerging esthetic sensibilities. It's as if only the ugliest and most grizzly toys can capture children's attention. Now the latest toys have shock or slime appeal, and eyeballs and entrails are equated with entertainment.

"Gross is in," as they say in toyland, and for the toymakers the object is gross profits. So we have Nose Ark, whose warty nose is so big it comes with a crutch; and Wired Wilma, with braces that conduct electricity; and Seezall Eggbrain, and Fat-So—all members of the Gross Out Gang. Now they're making fun of children themselves. In the process they're affirming kids' most antisocial behavior. If that isn't gross enough, there's the Mad Scientist Monster Lab with a vat so kids can sizzle the flesh off the bones of monsters. Or for a real hands-on experience they can Dissect an Alien with a scalpel. They can squeeze Mucous Pukous or Sinus Slimus to make "alien blood" pour from his mouth or nose. And if they can't fit the pieces together once they've cut them apart, there's a body bag in the kit.

*Mad Scientist Monster Lab*

The newest action toys go well beyond hand play. Recently toy companies have introduced child-size electronic battle gear. Remember how a game of cops and robbers or army was punctuated by "You missed"? No longer. Now, wearing helmets and shields, kids with "laser" guns can shoot it out and know they made a hit when the enemy's helmet lights up. Three hits and you're dead—no force field can save you! Worlds of Wonder's Lazer Tag was the hottest best-selling toy for boys in 1986. Inevitably there will be a big shoot-out since almost every major toy company has entered the laser-gun battle.

Again, the inspiration for Lazer Tag sprang from an entertainment concept for adults, not kids. The original game was designed for shoot-'em-ups between teams of college-age players. Translated to toyland, the technology may be seductive, but it's of questionable value for kids or their imagination. In effect, it takes aggressive play another notch closer to reality.

Similarly, the benign water gun has been updated with the Zap-It Liquidator. You know you've hit your mark when the colored liquid splats on your target's clothing. If you're lucky, he won't hit you before the "magic" stain dries and vanishes. And what do you do when the gun goes dry? Don't worry, just buy more refill cartridges.

Whether you approve of toy guns or not, you should know that as toy guns have become more realistic, the incidence of fatalities has risen. Most recently, a young teenager caught up in the excitment of a group laser-gun game was shot to death by policemen who mistook his "playful" stance and gun for the real thing. Despite the pleas of law-enforcement officials to ban the manufacture of realistic toy weapons, the arsenal and body count continue to grow.

Nor is violence limited to toy guns or the battlefields of action figures and superheroes. For preschoolers there are "gruesomely cute" plush toys—monsters with horns, handcuffs, chains, and cages. On TV, evil enemies chase even Rainbow Brite, Gummie Bears, and Wuzzles over rainbows and behind puffy clouds. It seems the preoccupation with good and evil can't be delivered too early or too often.

In their own defense, the animators point out that every show has a moral lesson. Not only do the good guys win, but the underlying evil is often summed up in a preachy sermonette as the show ends. It's something like the moral at the end of Aesop's fables. TV programmers are practically knocking kids on the head with positive messages. The messages are for the parents, the action is for the kids. Conveniently forgotten is the old chestnut "Actions speak louder than words," and the dominant actions on the screen are violent!

In effect, what we're experiencing is a kind of Gresham's law—the bad driving out the good. The violent, the ugly, and the saccharine have driven almost all other toys out of the market. It is the sheer quantity and sameness of both toys and shows that remains most disturbing, as well as the fact that both are being pitched to younger and more vulnerable children. It isn't the marketplace that speaks, but power marketing. And what the toy companies are marketing is an unending stream of violence, grossness, sweets, and uglies that pervades toyland and the playlives of children.

Like the toddler's interest in stomping through puddles, children's interest in war toys has traditionally been short-lived. But today the toy companies prolong the interest with their relentless marketing. As a result, it becomes more and more difficult for children to outgrow what should be a short phase.

For parents, the issue of violence raises all sorts of difficult questions. In an age of anxiety over nuclear destruction and terrorist attacks, what do kids get from toys and shows that resemble the six o'clock news? Some say the toys are a way for children to deal with their own fears. Others say that such toys and shows fan kids' anxiety and confirm their notion that conflict is best handled by force.

And what does it mean to kids when they watch shows with endless explosions, weapons, and chase scenes—violent shows on which no deaths occur? "There's no killing," parents like to say, and they're right. No blood or guts are spilled. Nowhere is the pain and suffering or the consequences of such acts shown. The forces of evil are never vanquished. Like Moriarity, they fade into the shadows, only to reappear and fight again.

But what do kids learn from such entertainment? Does it desensitize them? Some researchers believe that it does. Indeed, it may be that the appetite for ever more gruesome-looking toys is symptomatic of their blunted sensitivity. If the current cast of action figures looks like the stuff of nightmares to you, look again. Can it be that the quantity of such toys make them so commonplace as to be ordinary? Does the ability to manipulate and take charge of these symbolic forces of evil somehow help children in their own struggles with authority and rules and power?

In a single afternoon a child today watches as many "fairy tales" in one sitting as his father or grandfather may have heard over the entire course of childhood. Nor can today's electronic fairy tales be compared with the multiple and meaningful themes found in real folk literature. The young heroes and heroines of such stories deal with more than the forces of good and evil. In a sense, the classic fairy tales reflect the child's own struggle and journey toward independence. It is Jack who grows from a naïve boy to a brave young man who vanquishes the giant. Both Hansel and Gretel get rid of the wicked witch and find their way home. Such adventures present a meaningful metaphor in which the young hero is transformed not by an army of warriors but through his own kind of magic and power.

No issue in toyland has received more attention than

violence. Yet those who debate the issues do not even agree on a definition of what they mean by violence.

The toymakers tend cynically to practice denial in their definition. Hasbro's president recently said that GI Joe is not a war toy—"he's a defender of the peace." Similarly, the president of the Toy Manufacturers of America (TMA) claims that the only violent toys being sold today are guns, and that these represent only a minuscule share of toy sales. In this game of semantics the legions of armed warriors, their vehicles and grim settings are classified as fantasy.

At the other extreme, critics who evaluate TV shows tend to score every zap, screech, and bang as violence that is eating away at the moral fiber and mental health of today's children. Whether they're rating intergallactic robots, Ninja warriors, antiterrorist SWAT teams, or Elmer Fudd and Bugs Bunny, the incidents of violence are all rated on the same scorecard.

However, the lack of a definition does not diminish the problem. Indeed, if violence to the child's sensibilities is included in the definition, the problem is all the greater. Children are being assaulted by the tools of power marketing whether it's Rambo or My Little Pony.

This is not to say that the issue of violence is unimportant, nor that children's programming and toys would not benefit from more diversity, but rather, that the red-flag word "violence" is emotionally laden and diverts our attention from broader issues parents need to address.

If by some magic (or an act of Congress), violent programming as we know it today were to vanish, would the bigger issues go away? Probably not. In fact, even if there were a mandate that every program had to be about "love," the content and merchandise would change, but not necessarily improve. Parents would still have to deal with the pervasive lack of good taste, the simplistic and interchangeable story lines, the use of the sophisticated hard sell on easily manipulated and naïve children.

Viewed from that perspective, violence is just one of the many problems parents have to deal with in guiding their children and making choices on TV and in toyland. These are issues no amount of research can answer. These are questions parents will need to evaluate on the basis of the tastes and values they consider important.

Violence in programming influences toy buying from the preschool years on. Later in this book, as we look at toys for each developmental stage, we'll consider such toys as they relate to children's changing needs. We'll look at the research and at how parents are dealing with the pressure. Having viewed the toys in terms of ages and stages, parents will be better prepared to deal with issues as they arise.

Play, by its very nature, is an active pursuit. When the dominant play materials and themes originate in a passive medium, parents need to take a more active part in mediating the messages. Since toymakers don't know where to draw the line, parents will have to.

*Rambo*
*Coleco*

The influence of power marketing is so new it will be several years before research can assess its long-term impact. One can't help but wonder what would happen if commercial TV as we know it today were to vanish. Would the toy industry collapse? Would children play differently? Would they enjoy themselves more? Maybe. But romanticizing the past isn't going to alter the present. Change is charged with possibilities—some desirable, others less so. Newspapers, books, video, and TV are not intrinsically good or bad, but ripe with possibilities.

What can parents do? Well, in the best of all possible worlds, television would offer up a rich variety of programming to balance the current screech, splat, and zap scenarios. Activists are pressing for legislation to require stations to ban toy-driven cartoons, to present educational programming, and to reduce the level of violence. If these are issues of real concern to you, you may want to join forces with Action for Children's Television or the National

Coalition on Television Violence. But change comes slowly and children grow quickly. So what can you do in the meantime?

## *THE PARENT'S ROLE IN THE "BUY-ME" SYNDROME*

Since so many of children's toy requests are related to TV, we tend to think television is the sole cause of their seemingly insatiable appetite for toys. But putting all the blame on television is too simplistic. Parents don't buy toys simply because children ask for them. They buy toys for all sorts of reasons, some of which have little or nothing to do with the pressure from TV or the children themselves.

### *Great Expectations*

Toys are the gifts we give children for Christmas or Hanukkah, birthdays, and in-between times. Whatever else we may buy for holidays and birthdays, our gifts are almost certain to include toys. Some occasions call for toys; they're expected.

We not only buy toys for Christmas, we invite children to write lists and letters to Santa so that we are sure to get them what they expect. And we don't stop there. More often than not we like to add a few surprises—a touch of the *unexpected.* So we wrap up what they want, plus what we think they should want.

A recent study found that while preschoolers may receive as many as eleven or twelve toys on Christmas morning, the children themselves had requested only three or four. The findings suggest not only that parents are giving them more than they want or need, but also that they may be inflating their expectations. If a four-year-old gets twelve gifts, what should an eight-year-old expect? A giving binge may set a precedent for ever more lavish gifts in the years to come.

## *Gifts with a Guilt Edge: The Frequent-Flyer Syndrome*

Quite apart from holidays and birthdays, toys have also become tokens of guilt we carry in our luggage or attaché case. Like the cartoon husband who brings home flowers after a night on the town, busy parents often bring home toys as a kind of apology. It's a way of saying, "I'm sorry I'm so busy, but maybe this can make up for it."

Until recently these frequent-flyer gifts were brought home mainly by men whose work required travel and long nights at the office, or by the divorced father, who has traditionally been a soft touch at the toystore. But in today's world Mom is just as likely to be late for dinner as Dad. She may have a case that can't be adjourned for the second-grade play. She may have a sales meeting in Ohio the day of the school picnic. What she lacks in time she may hope to make up for with a little of that fabled "disposable income."

Given the high rate of divorce, single parents, and two-career families, toymakers may be the biggest benefactor of our changing lifestyles.

It's a safe bet that working mothers buy more "guilt-edged" toys than fathers of a generation ago did. Working is still an apologetic undertaking for women. They not only assuage their guilt with these consolation prizes, they also hope and expect the toys will keep the kids busy when they are at home. A busy child is not just a happy child, as the saying goes; no, a busy child needs less attention from Mom. Dr. Brian Sutton-Smith, an expert on children's play, has suggested that for parents the success of a toy is often measured by how well it occupies the child without adult assistance. We expect the toy to entertain and keep the child out of our hair. When the toy no longer holds interest, the quick solution is to buy a new toy.

While the frequent flyer may get some credit for trying,

all too often the costs keep escalating. A child's appetite for novelty can easily become insatiable. When parents who are frequently "on the road" (even if it's just downtown) come home with regular surprises, the pattern is established. A welcome home hug is accompanied with a "Whatcha bring me?"

Toys given as a substitute for parental time and attention are like junk food. They may provide instant gratification but not much nourishment. When we give children things instead of ourselves, what are we teaching them to value? When such gifts begin to represent the way we show our love, does it mean we don't love them if we don't bring a treat?

It's not that parents should never bring home a surprise or a treat, but such gifts shouldn't be made into rituals. Too many guilt-edged gifts can create a kind of emotional blackmail that feeds the "buy-me" syndrome.

## *Highway Bribery*

Much as we may disapprove, most of us have been driven at one time or another to bribe our children with rewards. "If you brush your teeth every day this week, I'll buy you a toy" or "After we go to the doctor, if you behave, we'll go to the toystore." Indeed, the doctor and dentist may have reward systems of badges or penny toys in their offices as well.

In a bitter custody battle, five-year-old Eric became a pawn in the weekly tug of war over weekend visitations. Although money (or the lack of it) was one of the fiery issues between his parents, there seemed to be no shortage of cash to spend on an arsenal of Transformers. Whenever Mommy bought a top-of-the-line robot, Daddy rushed out to buy one. Phone calls during the week became the arena for highway bribery. Neither parent seemed to trust that the boy would want to see them without the promise of a new toy.

While rewards can be useful, they are almost always more useful to parents than to children. We may start bribing young children with little gifts, but as they catch on, the ante goes up. Bribes can become a ritual that works in reverse. When the tables are turned, parents may hear their children saying, "I'll do this if you'll get me that." Clearly, bribery is another way that we ourselves fuel the "buy-me" syndrome.

## *Nothing's Too Good for My Child*

Even if we go to the toystore alone, we bring two people along: our adult selves and the tag-along child inside each of us. Sometimes these two selves see eye to eye, but often they are opposing forces.

Without realizing it, some parents see their children as extensions of themselves. Buying expensive and/or unlimited toys is a way of satisfying fantasies, both past and present. We may not have the money to buy a Maserati or a mink, but we can buy a remote control Alfa Romeo or a Cabbage Patch Mink Coat without much strain. If we can't have what we want, at least our kids can. Toys offer a kind of vicarious gratification.

Toy shopping also has a way of taking us back to our own childhood. We not only want to give our children what we had, we'd like to give them more. As a child you may have longed for a fully accessorized train set or a fantastic doll house. Now, at last, you can buy it. You've got the money (or credit card) and nothing's too good for your kids.

Buying expensive toys proves you don't need permission anymore. It's a way of proving to yourself and others how successful and powerful you are. Like the car you drive, the clothes you wear, and the house you live in, toys in great

> *A personal shopper sale [at F. A. O. Schwarz] of $1000 is no big deal . . . you start getting impressed around $5000. . . . I have one customer who has eleven children and comes in and does a separate personal shop for each one. . . . He just spent $1500 on his eleven-year-old daughter. And he has ten kids to go.*
>
> *Eloise Lipscomb,*
> **Wall Street Journal,**
> *December 18, 1986*

quantities and at any price have become another kind of status symbol. In a consumer-driven age, in which making and spending money is the way adults measure success, can kids be far behind?

All of us can get caught in this trap. If you've got expendable income, buying is painless and you can do it without stopping to think. Less affluent parents often go into hock trying to provide gifts beyond their means. "They're only kids once," they tell themselves. "I want my kid to be happy."

But, if parents don't control their own buying, what can they expect of their children? If you buy an eight-year-old a "toy" Mercedes for $7,500, what will he expect for high school graduation? Toys are the training ground for early consumerism and the first signs of a newly discovered malady known as "affluenza": Too often it's parents who are the carriers.

## *HANDLING THE "BUY-ME" SYNDROME*

As parents we need to recognize that when it comes to toys there's a lot of nonsense we need to deal with. It comes directly from the toymakers, from our kids, and from ourselves. Yet once we recognize the problems, there are steps we can take to define what we want—and what our children need.

### *Setting Limits*

The first and most obvious answer is to set limits on both TV viewing and TV-related buying. The earlier you set up the pattern of being a selective viewer and consumer, the better. Preselect the programs that are to be watched, and stick to it. Don't just turn the TV on and leave it on. Make a schedule and enforce it. Take the time to watch some of the cartoon shows with your child, even those you dislike. Discuss with your child the things that disturb you. Also talk about why you think there are better forms of entertainment. Better yet, take the time to introduce and sell your children on alternative ways of amusing themselves.

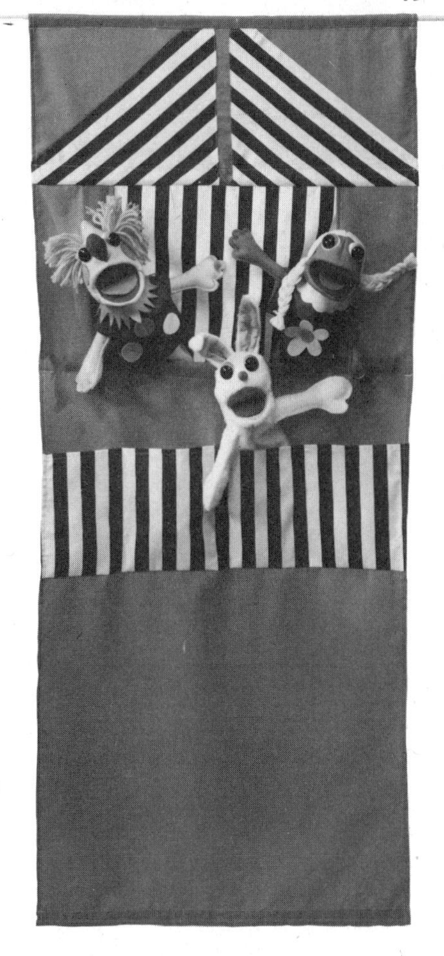

## *Expanding Horizons*

Remember that children's TV is not the only medium that can stimulate children's imagination and play. Books, videos, movies, music, computers, theater, art, and dance also provide stories that can spark play ideas. Competing with the instant access of TV may take some creative selling by parents, but the rewards may also lead to more creative forms of play.

## *Providing Variety*

There's an old story about a nursery-school child who painted every picture with black paint. His mother was alarmed. What was wrong? Was this symptomatic of an emotional problem? A little investigation revealed that black was the only color available on the easel. The same may be true of many children's toy chests today. We can't expect them to find multiple play modes if what they have is as limited as today's TV-related toys. Children need a variety of toys to exercise their muscles, their minds, and their imaginations. They may also need your example to get started. Open the paint and do a bit of dabbling on your own sheet of paper. Toss up a ball and think of the tricks to do with your hands before you catch it. Play a game of twenty questions while you're all clearing the table and stacking the dishwasher. Lend a hand with building props for a puppet show for the family. Few children will turn away from the double joys of a new game and a grown-up interested enough to play. Not even the best cartoon show can compete with such an offer.

*Puppets and Stage*
*Poppets*

## *Encouraging Inventiveness*

Parents can also limit the need for "more, more, more" by helping children find reasonable substitutes. "I buy the figures," one parent reports, "but not those thirty-to-forty-dollar environments. My son uses his blocks, Legos, and cardboard boxes to make castles, caves, and space stations." Children sometimes just need a little help in seeing alternatives or finding the raw materials. When parents give a bit of time and interest, their children's creative juices may really flow. In other words, buying less can end up being a way of supplying much more.

Parents don't need to feel powerless in the face of the choices. Limiting the quantity of viewing and buying is probably more realistic and, in the long run, more "educational" than a "never ever" or "anything goes" approach. You can express your feelings about toys and shows that don't meet your approval. You can stop a game that seems to be escalating and becoming dangerous or harmful. Without getting into shouting matches, you can talk about the issues. You don't have to shrug and surrender. You are in control but your long-term goal is to help your children forge their own controls.

## *All Licensed Toys Are Not Equal*

If you're a parent, you may limit your shopping to those select toystores that pride themselves on never selling licensed toys. On the other hand, keep in mind that not all licensed toys are bad. Parents need to look beyond the logo to evaluate the real play value in the toy.

The Cabbage Patch rage, for example, inspired an abundance of games and crafts with the Cabbage Patch logo. Many were standard kits for painting by number or tracing stencils. But one knitting kit offered equipment for making doll-size leg warmers, hats, and scarves. Here was a craft kit that expanded on the dolls' basic play value and encouraged a new skill.

Unfortunately, many of the licensed products lack much intrinsic play value. They may be decorative or satisfy the collector's bug, but do little to enhance play.

Remember, too, that many licenses don't last long. So a piece of equipment that might be used by several children over a period of years may get a dated look by the time your youngest child grows into it. Licenses also tend to be geared to one gender or another. Any boy or girl will ride a red trike, but a pink Cabbage Patch bike won't do if a little brother is next in line. Paying extra for the logo may be a luxury that ultimately means getting less value for your money.

Nor can you assume that all toys with the same licensed character are of equal quality. The owner of a license leases rights to many different manufacturers. Some produce better toys than others.

Parents may also fall into a trap with Sesame Street toys, assuming that all the toys the program has spawned are as educational as the program itself. But, of course, there's nothing more educational about Ernie's Rubber Duckie than any other rubber duckie. In 1985 there were close to sixteen thousand Sesame Street products. It's true that for Children's Television Workshop (CTW), the nonprofit organization that created Sesame Street, the revenues from Big Bird and his gang are a lifeline. With federal cutbacks in public television and education, CTW is more dependent than ever on returns from licensed goods.

Even so, the sheer quantity of Sesame Street products has glutted the market, helping to set in motion the wheels of the TV-toy connection, and with it, consumerism by the age of two or three. To a child of that age, does it matter that Big Bird was featured on an educational TV show before he turned up as a plush doll or a puppet, or on a talking phone, a bike, dishes, a coloring book, or finger paints? Probably not.

Like television, the licensing business is not going to go away. Although retailers are beginning to buy licensed merchandise with more caution since it is so faddish, parents and children will still need to look beyond the logo. A set of sturdy toy dishes with no logo may last longer than flimsy ones with a faddish decal. It's the quality of the plaything rather than the license that should determine what you buy and don't buy.

## All Toys Are Not Bought

When my grandchild Kate began to crawl, I hurried out to buy a rolling mirror toy that seemed exactly right. Kate never played with the toy, but the box it came in fascinated her. It opened and closed and was light enough to examine from every angle.

With young children particularly, we need to recognize the value of found playthings. All toys need not be bought. To a toddler, a shopping bag with empty plastic soda bottles is just as much fun to tote as a fill-and-dump plaything from toyland. If you're in the kitchen and can supervise, shove a chair near the sink. A drippy faucet and a few empty yogurt cups will keep your two-year-old busy longer than most toys. Kate's current favorite is her father's old computer keyboard with all those buttons to push.

Although many real things must be off-limits, those that children can use safely are often more compelling than store-bought imitations. They hold the extra attraction of being associated with Mommy or Daddy. They set the scene for using what's available. Whether your budget is tight or unlimited, before you buy, ask yourself: Is there something at home that could do the same thing?

## *All Gifts Are Not Equal; or, Gifts You'd Rather Do Without*

All of us have probably received a gift we could have lived without. But what do you do when well-meaning friends or relatives bring such gifts to your child? What if Grandpa comes calling with a GI Joe Aircraft Carrier for his only grandson? Or dear Aunt Steph shows up with a doll that eats, sleeps, talks, and even burps? This flow of toys from significant others can be as troubling as outside pressures to buy, buy, buy.

Unless the toy represents a real safety hazard, most parents surveyed found it best not to make a huge issue over such gifts. A few parents said they might hasten its disappearance after a few days. But experienced parents accept the fact that they can't control their child's world and keep it 100 percent pure. Rather than turning the gift into an issue and giving it undue importance, parents generally find it wiser to let the batteries wear out along with the child's interest.

Parents also say they make suggestions to relatives and friends before birthdays and holidays. Taking the guesswork out of shopping is often a welcome relief to gift-givers. To the extent that you can communicate your preferences, by all means do so. However, there's no need to agonize over every toy they receive. Like an occasional taste of junk food, it's not going to destroy them or the values you're trying to convey.

## *SAFETY AND FALSE ASSUMPTIONS*

If you've assumed that toys, like food and drugs, are subject to governmental safety testing before they reach the marketplace, you're operating on a false assumption. While the government does mandate specific toy safety standards, it provides no premarket agency or clearinghouse that tests or approves toys. Although many toy manufacturers do test their products, there is no law that requires them to do so. The toy companies are charged with the responsibility to comply with federal safety standards, and it is illegal for anyone to distribute toys that don't meet these standards. But it's not until accidents occur or consumers complain that the Consumer Product Safety Commission (CPSC) steps in to investigate suspect toys.

The CPSC determines whether any corrective action is to be taken, depending on the severity of the hazard a toy presents. According to the commission, there are approximately 150,000 different toys on the market and it is impossible to test them all. Of these, only about three hundred toys are tested by the CPSC, and half of these usually fail to meet government safety standards. The question is, what would happen if a thousand—or ten thousand—were tested? Better yet, shouldn't all of them be tested?

## *Accidents Related to Toys*

In 1985, seventy-eight toys and children's articles were recalled for violation of safety regulations or other hazards. Lead in paint, small parts that may be ingested or that may choke a child, sharp points and edges, long cords that can entangle and strangle, and battery connectors that overheat were the principal violations. The commission's latest report indicates that there were thirty-three fatalities associated with toys in 1984, and from hospital emergency room records we know that children suffered a staggering number of play-related injuries:

- 126,000 injuries related to toys
- 385,000 injuries related to bikes
- 65,000 injuries related to skates
- 23,000 injuries related to sleds
- 13,500 injuries related to skateboards

When we analyze the kinds of accidents related to toys, we find that the majority of injuries were cuts and bruises that children suffered by being hit by toys, or falling off, over, or into them. Riding toys such as trikes, rocking horses, and wagons accounted for more hospital emergency visits than any other toys. Other frequent offenders included disk-shaped flying toys, toy weapons such as guns, bows and arrows, slingshots, toy chests, and model cars and airplanes. Swallowing and choking on small toys or parts of toys, or sticking such objects into the nose or ears accounted for the next largest category of injuries.

## *The Parent's Role in Accident Prevention*

While the gap between safety standards and enforcement appears to need tightening, the life and death issues of safety can't be shifted solely to the toy industry or to the CPSC.

For instance, many accidents were related to toys given away at fast-food chains. So, the "freebie" wasn't chosen or evaluated with the kind of concern parents would bring to a toy purchase.

In the list of known fatalities, seven children died by choking on balloons. Five were electrocuted when their kites become entangled in power lines. Of the six children who lost their lives on riding toys, one rode into a pool and another was hit by a car. Another child choked on a marble and one strangled on a string attached to a toy whistle. With so many of these tragedies it becomes apparent that no amount of safety testing or regulation can replace supervision and knowledgeable toy selection. Here are some important things for parents to remember:

- *Supervision:* Nothing can replace your supervision. Children need to learn where they can safely ride trikes and fly kites. If a toy requires *adult assistance,* be there or don't buy it. Teach older children about the danger of small pieces and sharp edges on their toys that may be dangerous to younger members of the family. Don't count on them to remember. And don't rely on children of any age to use a toy just as it was meant to be used. Anticipate potential hazards and be watchful.
- *Select Toys with Safety in Mind:* Although many toys are age-labeled, some are not. Besides, many labels don't tell *why* a toy may be inappropriate or dangerous to a child. The recent death of two infants who strangled on crib gyms is a tragic case in point. Toys with strings, cords, ribbons, or elastic, such as pull toys, telephones, and crib toys should be of special concern with infants, toddlers, and even preschoolers.
- *Storage:* Provide open storage containers so toys aren't lying around getting stepped on or tripped over. Be sure that older children's toys with small parts are stored for safekeeping and remain inaccessible to younger children. Get rid of old toy chests with hinged lids that have spring-loaded lid supports. Year after year children are suffocated and strangled by toy chest lids that fall and trap them.
- *Maintenance:* Check old and new toys regularly. Repair or discard broken toys.
- *Brand Names:* More accidents seem to be related to toys from small and relatively unknown toy companies. A brand name is no guarantee, but there is a better bet that the well-known toy companies have done more safety testing.

- *Report Injuries:* If you suspect a product hazard or know of a product-related injury, report it to the U.S. Consumer Product Safety Commission. You can call their toll-free hot line: (800) 638-CPSC.

**SAFETY CHECKLIST**

- Look for sharp edges, hidden wires or pins, and nails that stick out.
- Check for size. Is the toy small enough to be swallowed?
- Inspect for parts that may pinch or clamp the child or entangle in child's hair or clothing.
- Make sure painted toys, or anything a child may put into his mouth, has a "nontoxic" label.
- Inspect for small detachable parts that could cause injury if swallowed or put in ears or nose.
- Check safety of battery-powered and electrically operated toys. Can child handle these safely?
- Check materials toy is made of. Is it easily breakable? Would broken parts be harmful?
- Avoid toys that shoot rockets, darts, or other projectiles with a force capable of causing injury.
- Avoid strings, cords, elastic, or ropes that may cause an injury.
- Look for toys that can be easily cleaned.

## *CONSUMERISM 101*

It should be some comfort to know that children's intense fascination with TV-toys is usually short-lived. For some children, the programming and the toys simply become boring. Others become disenchanted when toys they see advertised don't live up to their expectations. In the Bank Street Survey, most children up to the age of seven, the so-called age of reason, loved not just the toys but the commercials. But by seven and a half, children became somewhat critical about both the toys and the ads.

"I don't trust them," a seven-and-a-half-year-old boy wrote. "Some commercials are false and they rip you off!" An eight-year-old wrote, "They don't convince me!" By nine, children wrote, "They're stupid! Stupid and boring!" "I *hate* them because they make things look better than in real life!" "They make toys look better than they are." "They fall apart!"

One of the most telling tales came from a first-grader who told me: "I want a Rainbow Brite doll, but my mom says no. She hates Rainbow Brite because she says it's old-fashioned —you know, a sissy toy from when girls had to grow up to be mommies or teachers or stuff like that. But I think she's wrong. I want Rainbow Brite because her song on TV says she can make you happy even when you feel sad."

Clearly, Jenny and her mom were operating on different wavelengths. In the end the commercial message both won and lost. Jenny got her Rainbow Brite and a few weeks later confessed, "My grandma bought it for me and I like Rainbow Brite—a lot. But it's not true. She doesn't make me happy when I feel sad; she doesn't brighten up my day. That's just a song."

Occasionally, there's genuine value in buying what you may consider a "bad" toy. When children have some firsthand experience with toys that don't live up to the claims made in ads, toys that are essentially boring or flimsy or inspired by fads, they learn a small but significant lesson about consumerism. This is particularly effective with older children who frequently have birthday, holiday, or allowance money to spend. Learning to think twice is seldom learned quickly. But as one mother explained, "When my ten-year-old wanted a big TV-toy that was obviously a piece of junk, I said I would not spend my money on it, but he could buy it. After he had saved for weeks, the toy broke in the first hour and I didn't have to say anything. I certainly didn't say 'I told you so.' I was able to be sympathetic and he was able to be angry about being ripped off."

In teaching children about consumerism, don't overlook the positive value of a firm "No." Since television pumps and primes their nag power, children need to learn that nobody gets everything. Toys are just the beginning of a long list of "buy-me's." A few years up the road there will be records, comics, clothes, cosmetics, jewelry, and "equipment." Learning to live in a world full of things one can want but not always own is part of growing up, and a lesson to be learned and relearned, even as an adult.

## *LEARNING TO CHOOSE*

Toys, in and of themselves, don't teach values—parents do. Children learn values from the ways the important people in their lives treat them and each other. We send messages directly and indirectly: values are transmitted by what we say or don't say, by what we do or don't do.

Young children generally accept what Mommy and Daddy believe as their own beliefs. They may not always want to do what we say, but when in doubt parents are quoted (and misquoted) as the ultimate authority. During the school years other values begin to reinforce or contradict those initial beliefs. Developmentally, it is a long journey to that place where all the threads are woven together to form the fabric of one's own beliefs.

While we may wish we could shut the door and keep the world out, we know that's not really possible. In fact, it may not even be desirable. How can we teach children to live in a consumer society if we keep them hermetically sealed off? At what age will they be ready for release? And how will they learn about the concept of choice without learning to exercise choices themselves?

Although we know that children learn best about the nature of objects and people from direct interactions, these are by no means the only experiences available to them. Television allows them to travel beyond the confines of their immediate neighborhood without setting foot outdoors. With VCRs, they can watch an event not only as it happens but also in an instant replay. And with home computers they can play chess or land a plane at O'Hare or create a new persona for an adventure game. All these simulations are possible without any human contact. In this fashion, technology and the media expand the world at the same time that they isolate children from others.

Such simulated experiences are another piece of the learning agenda our children need to grapple with. For parents, this means having to clarify any confusions that may arise. Most of the children who watched Peter Pan in the 1950s did not try to fly. Most children today who watch cartoons know that if they hurt a friend he won't bounce back unharmed. It can be useful for parents to see the language of simulation, its conceits and motifs, as another symbol system they need to help children learn and understand. At the same time, they can provide a balance by offering their children other kinds of materials and stories as springboards to play and imagination.

In a real sense, parents can use the media and toys as a catalyst for conveying values. Rather than shutting the world out, we can respond to those things we find crass, commercial, vulgar, or violent; we can discuss and clarify what we find objectionable or distasteful. It's from their parents that children learn some early lessons in becoming selective consumers and media users. Who else can teach them that we don't buy everything we see or want, not even in the bookstore, record shop, or toystore?

At every stage of development, parents have opportunities to transmit their most deeply felt values and tastes. It's not just what we say but what we do that has lasting impact.

If at the end of the day our chief form of entertainment is an easy chair and a chase scene on TV, the message is not lost on our children.

It is reassuring to keep in mind that our likes and dislikes in behavior, food, clothing, books, music, TV, toys, and all manner of things affect the texture of our children's day-to-day experiences. In the many ways we express our pleasure and displeasure, our disappointments and enthusiasms, our concern and criticism, our approval and disapproval, we pass on a set of values that usually runs longer and stronger than most outside messages.

As parents, it is our job to weigh the choices. We have the power to say yes or no. And it is our responsibility to do so.

TWO

# *Toys by Ages and Stages*

# II. *Toys for Infants: The First Year*

Gone are the days of the cool, calm, pastel nursery. Today some babies are being stimulated to distraction. For many children, the toy glut begins the moment they are born.

Although adults often worry that too much attention may "spoil" a baby, few parents seem to have the same concern about too many toys. Yet, a few years later these same parents are appalled by their child's appetite for toys. It may well be that the wheels of the "buy-me" syndrome are set in motion during the first year of a child's life.

## *RESEARCH AND THE TOYMAKERS*

Researchers have told us that infants respond to a stimulating environment, and toymakers have dutifully translated these findings into an abundance of products to dangle from crib, carriage, changing board, high chair and carseat. But they haven't stopped there. Crib sheets, bumpers, stretch suits, blankets, and even diapers are now available with patterns. So is wallpaper. And just in case baby can't sleep, you can buy plush dolls with sound effects that simulate Mother's heartbeat or the swishing sound of amniotic fluid.

The notion of constant artificial stimulation is being sold as a valuable way to communicate with and comfort babies. As a result, many newborns begin life in an atmosphere that can only be described as frenetic.

It's not that toys to comfort or stimulate are bad or without value. Not at all. But there are no "things" more interesting, stimulating, or important during a child's early months than a real live person. People are the most versatile and fascinating playthings in a baby's world. Objects are not a substitute for people, and a young baby can't interact with objects without the assistance of people.

The problem is not with stimulation, but with the sheer quantity of it. Remember, researchers look at one stimulus at a time. But translated into merchandise, the multitude of stimuli offered in a typical nursery may amount to a bombardment of the senses. Of course, babies are wonderfully adept at turning away when they've had their fill. So, parents need to be sensitive to their baby's interest or lack of interest—knowing when to stop as well as when to begin. Indeed, an overload of stimuli may be more of a stressful distraction than an attraction. Imagine yourself trying to watch a film on TV, listening to a tape, and talking on the phone—all at the same time. It can be done, but it's not much fun.

### *Avoiding Too Much Too Soon*

During this first year it's easy to be seduced by the novelty of buying toys. You'll need some, but not many. Save the money—you'll need it for college tuition!

Often the overload in the nursery builds like clutter on a coffee table. But infants, like all of us, begin to ignore things that are always in sight, although they will take note of the new and novel. The solution is to change toys from time to time rather than always adding new toys. Providing fewer things at a time will capture their interest far more and may also help them focus their attention.

## *FIRST TOYS: BIRTH TO SIX MONTHS*

Less than five years ago a survey showed that 70 percent of mothers in the United States believed that newborn infants could not see. The fact is, infants come into the world with a full set of sensory equipment. Not only can they see, they can hear, smell, taste, and touch. It's through their senses that babies learn about people, things, and themselves.

Even before babies can reach out and take hold of things, they begin to explore the world with their eyes and ears. But until they are about three months old, infants can focus their eyes only within a limited range. As a result, their toys need to be eight to fourteen inches from their eyes. Objects that move slowly and produce a gentle sound will capture their attention more than a fixed object. A musical mobile on the crib rail will attract more attention than a static picture on the wall. Studies indicate that infants under three months look to the right 80 percent of the time, so a crib toy suspended overhead or on the left will be of limited value during your child's early weeks.

## *Mobiles*

A musical mobile attached to the side of the crib rail adds interesting sound and action to baby's horizontal view of the world. At first, you will have to make the mobile move. Eventually, baby will discover that by kicking her legs or moving her arms she can make the mobile move herself. Her actions will create a reaction—an early lesson in cause and effect. She doesn't have to touch the mobile to make it move. Just moving in her crib will do that. Of course, for safety's sake the mobile should not be within baby's reach. As baby matures and begins to bat at and grasp objects, the mobile should be suspended higher and eventually removed. Remember, many crib toys are intended only for the first five months. They may have small parts or strings that can be a hazard to older babies who tend to mouth anything they can grasp.

In selecting a mobile, consider how it looks from baby's point of view. A mobile with small pastel pieces may seem pretty to you but in fact is less interesting for baby to gaze at. Infants are more attracted to bright primary colors, objects with sharp contrasts and clear but simple features. Fisher-Price's Dancing Animal Music Box Mobile is well designed so that baby can see the whole object rather than just the bottom edge.

To provide new interest from time to time you may want to move the mobile from the right crib rail to the left as baby begins to scan more of the world. Since mobiles are expensive and of short-term interest many child-care books recommend homemade mobiles suspended on wire hangers. But in the interest of safety, particularly if there is another child running about, such homemade inventions are best avoided. Buy or borrow a mobile that fastens securely or attach toys for gazing to the crib rail. They won't move, but they won't fall on the baby either.

***Warning: Do not use elastic to attach toys to crib rail!***

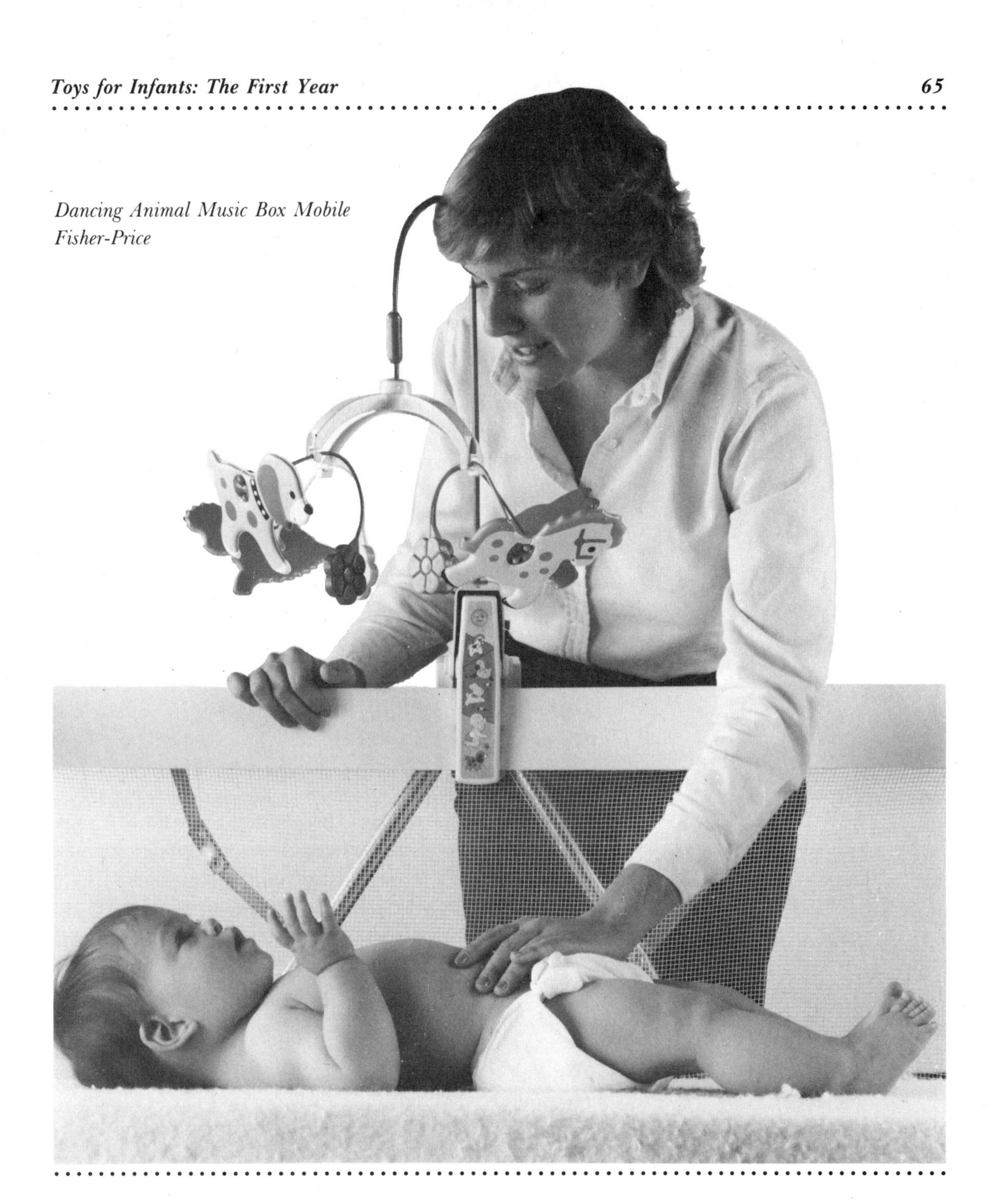

*Dancing Animal Music Box Mobile*
*Fisher-Price*

## *Mirrors*

Also interesting are the ever-changing images, movement, and light reflected in a mirror. Attached to the crib rail or beside the changing table, a nonbreakable mirror made of polished metal is both safe and fascinating. Although babies don't recognize their own image until later in their first year, they will begin smiling at their own image by three months. Johnson and Johnson's Baby Mirror has a smooth plastic frame with round edges and a screw mount for attaching the mirror securely to the crib rail. Later it converts to a stand-up mirror for tabletop.

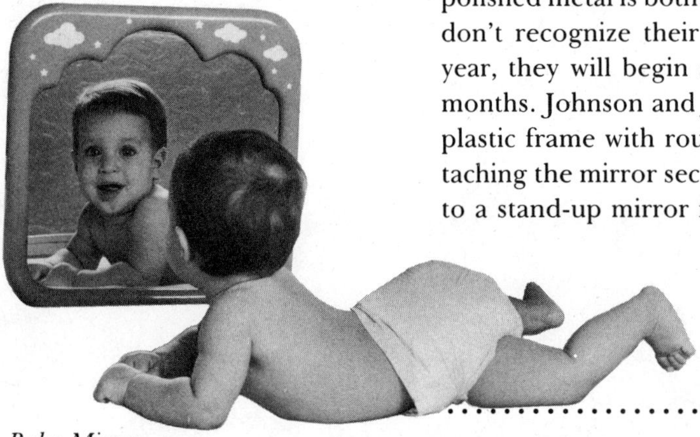

*Baby Mirror*
*Johnson and Johnson*

## *Lap Toys*

Between feedings, changings, and naps, you and your baby will enjoy some playthings for lap time. During the first three months you are your baby's first and best plaything. No toy is as fascinating as your smiling face, your soft voice and your gentle hands. Indeed, many playthings for new babies are really more for grown-ups than babies. They serve as props for early social interactions when everyone is getting acquainted. They also encourage baby to visually track objects and begin batting and reaching out. To begin with, you're the one who will have to lead the play and choose the playthings. It's through you that baby will discover the world of things.

Squeak toys, rattles, teethers and soft cloth toys all offer new sounds and sights that will attract baby's attention. Toys that are brightly colored and have interesting sound and visual effects will be more interesting for baby to bat at than a solid colored object with no sound. These early toys will eventually be played with independently, examined from every angle, tasted, passed from one hand to the other, banged, twisted, and turned. So select washable toys that are safe enough for baby to play with alone.

You'll notice that if you put a small rattle in baby's hand, she'll grasp it reflexively. But, she'll also drop it the same way, with no effort to see where it fell. For now, what's out of sight is also out of mind. She'll be three to four months old before she can purposefully reach out and take hold of a rattle or teething ring. Even then, many rattles are either too heavy or too small for baby to manipulate safely. In fact, small rattles that often come with flowers and gifts for baby can be downright hazardous!

According to the Consumer Product Safety Commission, in 1985 fourteen children choked to death on baby rattles. The oldest baby, only eleven months old, fell with a rattle, causing it to be jammed in the throat. Accidents were also reported when infants partially swallowed rattles while sucking on them. The commission reports indicate that similar choking tragedies have occurred with squeeze toys and teethers. A baby's mouth is very flexible and can stretch to admit shapes larger than you might imagine. Accidents have also been reported when handles on rattles were small enough to lodge in a baby's throat and block the airway. Some of the rattles involved may have been included as cheap decorations on packages, but costly antiques and modern silver rattles and teethers with small pieces can be deadly, too.

Soft and washable, Kindergund's Linky Dinks are three brightly colored shapes covered in soft velour. They're easy to grasp and make a jingly sound. Equally interesting to the touch are small terry cloth creatures with stitched features made by several manufacturers such as Gund, Dakin, and Kiddicraft.

*Consumer Product Safety Commission's recommended minimum rattle size*

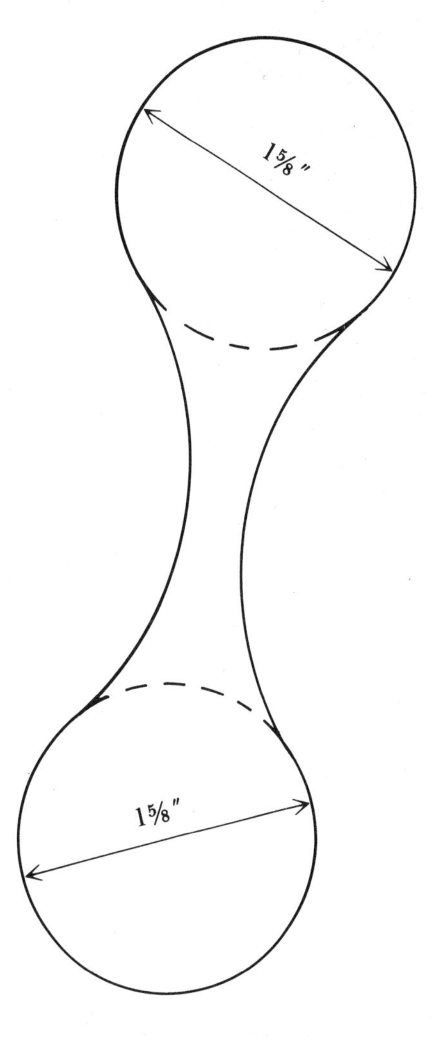

*Kindergund Clown Rattle*
*Gund*

Also interesting for getting baby's attention is the Animal Grabber Puppet from Fisher-Price. Slip your hand inside and move it from left to right in baby's sight line. Once baby learns to track the puppet from side to side, move it up and down or around.

Babies don't need toys strapped to their wrists or feet in order to stimulate body awareness. A baby's own fingers and toes provide enough fascination and stimulation without add-ons.

## Crib Gyms

During the third month, baby typically discovers one of the best toys in the world—his own hands. He may spend a great deal of time watching those fascinating fingers that are no longer clenched in a fist. He can bring his hands together and reach out to grasp a toy—a bit crudely, granted, but with more refinement than before. Often this is the time when a crib gym is introduced. But keep in mind that crib gyms are of questionable play value. In fact, some are downright dangerous. Let's look at the safety problems first.

- Crib gyms are intended for horizontal babies. They are not for pulling up on.
- Crib gyms should be removed by five months or when baby can push up on his hands and knees.
- A crib gym should never be attached to crib rails or playpen with elastic. Read and follow manufacturers' directions carefully. Elastic stretches and the toy can entrap and strangle the infant.
- Don't borrow or use an old crib gym that may not meet today's safety standards.

Although crib gyms have been considered standard gear for years, the potential for tragedy is real. Elasticized crib toys like Johnson and Johnson's Piglets have been recalled, but other similar products may still turn up. While some toy companies mark the product itself with safety cautions, many safety and age labels are only on the disposable packaging, so parents install the toy and forget. Indeed, many of the victims of strangulation deaths from crib toys were able to sit up and even stand.

In other words, they were well past the five-month age limit. But it's not just age that counts. Some babies can push or pull up before five months. One baby became entrapped when the sleeve of his sweater got entangled in the crib gym.

Safety aside, many crib gyms are more frustrating than stimulating. Since baby's grasp is crude, dangling toys are hard to grab hold of. Indeed, many store-bought and homemade versions are basically untouchable since the objects dangle and swing out of baby's grasp. Since babies generally go from grasping to mouthing and shaking the objects they can hold, several manipulation toys may be of more use and interest.

## *Choosing Soft Toys for Infants; or, All Bears Are Not Equal*

Newborns often receive a menagerie of soft toys as gifts. Buying a bear or a bunny is more fun than picking out a stretch suit. Besides, I suspect that everyone is hoping to give the huggable that becomes baby's favorite. Since older babies often do form an attachment to a favorite animal, it's especially important that the soft toys chosen now can go the distance.

Not all bears or bunnies are created equal, especially not for babies! A large percentage of the toys recalled because of safety hazards in 1985 happened to be plush toys! Soft toys with plastic eyes and snouts are potential choking hazards to infants. Indeed, any frills or features such as bells, buttons, flowers, whiskers, ribbons, and yarn wigs should be considered what lawyers call an "attractive nuisance." Such doodads are the very things a baby will mouth and possibly choke on.

True, the newborn can't reach out and take hold of things, but why put a doll in the crib that can become a hazard in a few months? Musical plush toys may seem soothing for a newborn but not when baby can roll over and get jabbed by the metal wind-up key. The music box on the mobile or tied to the crib rail is a better choice.

Handmade toys may be made with love, but that doesn't make them safe. Before choosing a charming calico cat at a craft fair, ask about the contents and check the seams. In general, whether you're buying a handmade toy or a store-bought one, look for animals or dolls with stitched-on features that can't be chewed instead of plastic snouts and eyes. Floppy ears and tails are safer than ones stiffened with wires that can poke their way through fabric and jab. Steer clear of oversized dolls that are too bulky for baby to explore, and don't be seduced by cute but cheaply made novelties. Good plush toys are not cheap, but a well-made doll or animal will be enjoyed for years.

For this first doll you'll want to find one with clear and contrasting features. It should be lightweight and not too big or heavy. Often in the second half of the first year this doll may become a constant crib-time companion, a comfort like Linus's security blanket. With that in mind, you'll want to be sure that it has no hidden prickles. Good choices for the long haul are Gund's Stitch Bear or the slightly smaller Kindergund Bearbaby. Both are totally soft and safe teddy bears with embroidered features. Or look for Playskool's Blankies, made of soft blanket fabric trimmed with satin and stitched features.

What do you do with the menagerie that well-meaning gift-givers send? Put them away. In a year or two, when baby is past the stage of putting everything in her mouth, these will be welcome playthings.

OPPOSITE
*Stitch Bear*
*Gund*

## *Have Toys, Will Travel*

With all the basic paraphernalia that are "must-haves" for mobile babies, you'll want to keep the take-along toys to a minimum. Nevertheless, a few favorites for chomping, tossing, shaking, and banging make useful distractions when the going gets tough. A long car ride or flight, or an unfamiliar hotel room can be made a bit more comfortable by bringing along a few touches from home.

Playskool's Fold 'n Go Activity Quilt provides an easy-to-tote surface for precrawling babies to rest on. It has several toys attached along one side so baby can stretch out in comfort instead of lying on top of the toys. Ideally, an activity quilt ought to reverse to a vinyl surface for easy changes. Unfortunately none has that feature.

If a music box is part of the soothing ritual for bedtime or naps, take along a small musical toy to tie onto the crib or carriage.

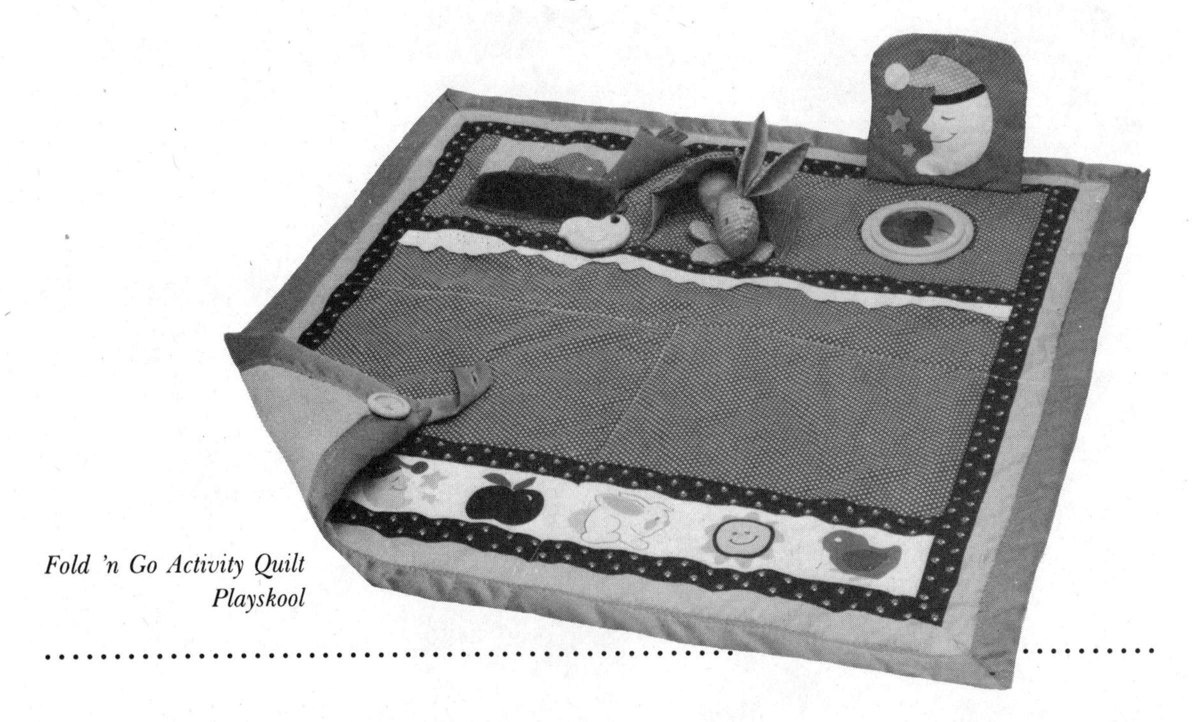

*Fold 'n Go Activity Quilt*
*Playskool*

Although there are dozens of toys for carriages, strollers, and car seats, few hold much long-term interest and some are downright dangerous. Steer clear of traditional elasticized toys that stretch across the carriage or stroller. As with crib gyms, they can be a strangulation hazard. Even dangling carriage toys suspended on multiple strings of elastic can be deadly. Ornaments for carriages and strollers are usually more decorative than entertaining. With a young baby they're just excess baggage and for the older baby they're boring. By the time baby is sitting up, few toys will be more interesting than the ever-changing passing scene. To keep toys in safe but easy reach, look for manipulatives that can be fastened to the stroller or car seat with a plastic ring. These can be used in transit or as a floor toy at home.

## *SIX MONTHS TO TWELVE MONTHS*

Once baby begins sitting up, he has a new view of the world and develops a fascination with manipulating objects. He explores toys from every angle with eyes, fingers, and mouth. Using two hands he will bang two toys together or spend a long time passing a toy from one hand to the other. The look on his face should tell you this is serious work, a way of discovering the nature of objects and himself. More than ever, he is interested in making things happen.

### *Manipulative Toys*

Standard equipment in most nurseries is an activity board attached to crib rails. Some child-care experts and parents claim that activity boards are basically boring. Nevertheless, most of the parents interviewed in our survey indicated that the multiple action boards are enjoyed not only at this stage but well into toddlerhood, when they become floor toys.

*Red Rings Rattle*
*Johnson and Johnson*

*Activity Center*
*Fisher-Price*

Activity toys with doors to open, buttons to push, wheels that spin, and knobs to turn offer babies the opportunity to practice fine motor activities with payoffs they can see and hear. If you buy one, make sure that the activity board you choose has parts that are easy to move, or the toy may be frustrating.

Out of the crib and seated on a blanket on the floor, babies enjoy slightly different types of toys. In fact, multiple toys are generally of more interest than activity boards at this stage. An easy-to-grasp baby mirror offers ever-changing images to look at. A ring with multicolored keys or discs is great for chomping and shaking or fingering. A set of bright shapes such as Fisher-Price's Baby's First Blocks is interesting to look at and easy to grasp, bang together, shake, and release. They come in a shape sorter container. Put the lid away for now. Stick to filling and dumping games. Or consider ImaginAction's Block Sensations, six transparent cubes, each filled with a different sensory element to feel, hear, see, or smell. For a change of texture, fabric-covered foam blocks are fun to throw and roll. Also perfect for pulling apart, throwing, and dumping into boxes are brightly colored pop beads.

*ImaginAction's Block Sensations Worlds of Wonder*

*Mirror*
*Kiddicraft/International Playthings*

*Snap Beads*
*Lardy/International Playthings*

*Bruna Fabric Blocks*
*International Playthings*

*My First Car*
*Ambi*

In selecting manipulation toys for this age, look for toys that can be used in several different ways. Johnson and Johnson's Red Rings or their Star Rings Rattles with multiple stars to grasp and turn are more interesting to handle or chomp on than a plain round teething ring. Fisher-Price's Flower Rattle has a mirror on one side, a face on the other, petals to teethe on, and sound effects that make it more interesting than a simple rattle. Ambi's Twin Faces and Kiddicraft's see-through teething ring both offer sound, motion, and chomp appeal. Remember that babies need toys that give them variety as well as quick and easy feedback. Often the variety can be built right into the toys.

As babies themselves become more mobile and begin crawling, toys that move hold new interest. A small but simple toy car with sturdy wheels is fun to push back and forth. Ambi's My First Car and Playskool's Wee Wheels are the right scale. So are Johnson and Johnson's Rhythm Rollers, with chime-filled barrels that can be stacked or loaded on a board. Toys placed slightly out of reach provide motivation for crawling. Remember, however, that your object is not to tease but to tempt.

## *Balls*

Few toys lend themselves to as much active social fun as balls.

For beginners who can't crawl, a slick vinyl ball that rolls too easily may be frustrating. It may also pack a wallop when it's tossed. A simple, lightweight, fabric ball will be a basic toy for now and months to come. Dakin's Plush Chime Ball is brightly colored, lightweight, makes a jingly sound, and has a strap to grab onto.

This is the stage when chime balls such as Playskool's Chime Bird, which can't roll out of reach, are of interest.

Keep in mind, though, that weighted chime balls are wonderful at this age but may become lethal weapon's for ones and twos (they think they are balls and throw them!). Also nice for easy grasping is the Touch 'Em Clutch Ball from the same company. Made of fabric, it has six textured handles and a jinglebell sewn inside for sound appeal. Or consider Learning Curves Curiosity Ball, a fabric-covered foam ball that opens up to reveal four pockets with activity toys inside. Though no baby needs them all, each of these balls provides quite a different experience from the others.

***Warning: Ping-Pong balls, which fit into baby's mouth, are too small for safety. Soft spongeballs may also be hazardous if chomped on or picked at.***

Once baby begins to crawl, an inexpensive beach ball, slightly underinflated, is light and easy to catch up with.

*Balls in a Bowl*
*Johnson and Johnson*

*Rhythm Rollers*
*Johnson and Johnson*

## *Independence and Interdependence*

Given a few bright and interesting objects to manipulate, baby may be content to play independently for a short time. Rather than plunking down everything at once, put five or six objects in a box or bag. This is the stage when filling and dumping objects is part of the game plan. When these objects become boring, change the scenery. Don't leave the old props lying about. Changing the toy supply keeps alive the element of surprise.

Of course, even an endless supply of toys will not keep babies busy for long stretches of time. This is a good time to establish a "playtime" together, quite apart from the time you spend playing while dressing, feeding, or bathing your baby. You don't need to make it a formal occasion, but set some time aside for one-on-one play. Turn on a tape, pick up your baby, and dance together; or get a pot and two spoons and beat out a rhythm together.

If you haven't started reading aloud to your child, now's a good time to start. Look for cloth books or those printed on sturdy cardboard stock. Though they're not much interested in stories, babies love the sound of your voice, the rhythm of your words and the comfort of your lap. Sharing books at this stage not only encourages language development, it can be a surprisingly wonderful experience for you and your child. After sharing a book, your baby will come to love the "mechanics" of turning pages and examining books independently. Small chubby books are easier for little hands to explore. Magazines are also great for leafing through, looking at, and often ripping to bits. Since torn paper will be mouthed, you'll need to be there.

Often interest in a toy is extended when parents take the time to "model" little games.

Aside from your usual games of patty cake, peekaboo, or "this little piggy went to market," try some of the following:

- Play some hide-and-seek games with objects. Put a toy in a box, under a diaper, or behind your back. If you "hide" the object while baby is looking he'll "find" it very quickly.
- Stack a few soft blocks and show baby how to knock them down. Do it again.
- Drop pop beads one at a time into a metal mixing bowl or tin canister. Baby will enjoy the sound as well as the action. Dump them and he'll try to do it again.
- Use baby's teddy bear to play patty cake or peekaboo or ring-around-a-rosy.

These needn't be formalized "lessons," but little games your baby will enjoy. In time he'll initiate the action and even indicate the playthings he prefers.

Following baby's lead is part of the dialogue of taking turns and playing together. If a toy or book doesn't interest him, don't push it. Be flexible and try something else. A jack-in-the-box may produce giggles or a pucker followed by tears.

Sometimes a toy that is uninteresting today becomes a favorite tomorrow. The success of a toy is less important than keeping the fun in playing together. If it's not fun, it's simply not play.

Once they begin to sit up in a high chair babies inevitably begin the classic game of "I drop it—you pick it up." It may become irritating for you, but for him this drop-and-pick-up game marks a developmental landmark. Baby is not only exercising you, he is exercising new kinds of intellectual and physical skills. Now that he can release an object intentionally rather than accidentally, he begins to watch where it goes. Out of sight is no longer out of mind. While a suction-based rattle may be more convenient for you and of some short-term interest to him, it should not replace the fun and learning of the drop-and-pick-up game.

## *Being Resourceful Is Playful*

Many household objects are as interesting to your baby as toys from the store. Plastic containers, boxes, paper, pots, spoons, and juice or coffee cans with smooth edges offer plenty of fascination for banging, rolling, scrunching up, twisting, and turning. A paper bag or shoe box to fill and empty has plenty of play potential.

Not all games at this stage involve toys. You don't need anything but willingness to play a game of peekaboo. Or put baby on your leg and give her a ride to the rhythms of a nursery rhyme. Finger games or a happy game of patty cake are sure pleasers. Once your baby is on her feet, try a gentle game of ring-around-a-rosy. You don't need to circle, just sway and say the rhyme. When you get to the part that goes "all fall down," do it! It's a perfect way of practicing the inevitable art of falling, a skill all new walkers need.

Singing, talking, making funny faces, and active but gentle roughhousing are all part of the repertoire of parenting. Indeed, whether you're changing diapers or fixing lunch, a bit of playfulness can make routine activities into pleasant interactions for you and your baby.

## *Playpens and Walkers*

During these early months when babies are learning about learning, nothing is more basic than their natural curiosity. As a result, equipment for babies that confines them to any one place for extended periods of time is generally a poor choice. That includes cribs or strollers as well as other equipment. A crib or playpen full of toys is no substitute for the freedom to explore.

However, there are times when a baby needs to be put down in a safe place. On a limited basis a playpen may be a useful piece of equipment. For the baby who's learning to pull herself upright and step sideways, the playpen with a pad may be safer and softer than a bare floor. When chores need to be done, the playpen may be a safe haven if it is set up in a room where baby can see and "talk" to others. It's certainly better than being packed off in a crib alone.

Walkers also offer the pre-walker a chance to use her feet to get about and be more sociable. For starters, baby will use her feet and propel herself backwards. With a little practice she'll soon get the hang of it and get where she wants to go. However, walkers do pose several potential safety hazards. Some models, particularly older ones, lack safety bumpers or have collapsible legs that can entrap small fingers. For that reason, forget about old walkers from the attic or from your well-meaning friend or neighbor. A well-made walker is constructed with round bumpers to protect baby from tipping or slamming into walls. They come equipped with trays that can hold toys for exploring. Since babies should never be left alone or unattended in a walker, think of it as a toy that baby will enjoy with your close supervision. And remember that you'll need to prepare the area so that your baby can't run over obstacles or steps.

***Caution: More than fifteen thousand children are injured each year while using baby walkers. Supervision is a must.***

## Bathtime

At this age your baby doesn't need too many toys for the bath. Many household objects work as well and are just as fascinating. A washcloth, a sponge, a set of cups for pouring —all are fine. The object is to make bathtime fun rather than a hassle. A few bath toys may add to the pleasure. Johnson and Johnson's Balls in a Bowl can be used in the tub and out. Playskool's Tub Pals come with three little sailors that float in or out of their vessels. Or look for Kiddicraft's Mother Duck with three small ducklings that can be loaded on her back. Many of the more elaborate water toys will be enjoyed more later.

*Tub Pals*
*Playskool*

## LOOKING AHEAD

As your baby gains greater mobility, some of the toys she enjoyed earlier will be used in new ways. Others will be packed up or passed along to a baby cousin or neighbor. At every age parents need to keep tabs on toys that are outgrown, broken, or just taking up space. Like clothes that don't fit, too many toys in the nursery will simply get in the way.

By the end of your baby's first year, the mobile, the squeeze toys, the rattles, and most simple manipulative toys have been outgrown. However, baby is likely to be babbling to a well-loved and slightly scruffy looking teddy bear. She may be dumping pop beads in a pot or trying to fit them together. She may be the one to launch a game of patty-cake or catch. If she is up on her feet early, some of the walking toys in the next chapter may be of use.

No baby needs all the toys in this chapter. Whatever you buy or make during this first year, remember that toys are just the props of play, not a substitute for the people who play with babies and enjoy it!

# III. *Toys for Young Toddlers: The Second Year*

## *THE TODDLER'S WORLD*

Up on his feet and ready to go, the whole world is the toddler's plaything—or at least he'd like it to be. With his newfound mobility anything that's within reach is worthy of investigation. Toys are by no means the most interesting objects to be explored, although in a sense, they may be a safer and more reasonable substitute for the multitude of real things he'd love to grab, turn, twist, bang, lug, throw, taste, and test.

As he makes the transition from wobbly first steps to sure-footed walking, and ultimately to running, your toddler is a study in motion. Few toys can compete with the challenge of getting about. Steps are there to be climbed, cabinets for opening and emptying, cords and tablecloths for reaching and pulling.

Though your toddler spends a lot of time listening and watching, he's no longer content with simple visual explorations. Now he wants to experience his expanded world with all his senses. But in his eagerness to get from one place to another, his sense of curiosity far outweighs his ability to predict outcomes or foresee danger.

## *Setting the Stage for Learning*

In providing a stimulating environment that encourages both mobility and curiosity, parents need to set the stage for freedom to learn and play. Moving medicines, household cleaners, and other dangerous substances out of reach is just the begining of childproofing.

It's not just for the child's safety or the preservation of family heirlooms that experts advise childproofing. Clearing the decks for play means limiting the number of no's you'll have to use to punctuate the day. Being able to say yes is an excellent way of reinforcing your child's best resource for learning—his or her curiosity. Rather than setting up yourself and your toddler for endless confrontations, put the "untouchables" out of sight. This isn't the time for lessons in "don't touch." Touching is what toddlers do. It's the way they learn.

Young children learn less from long-winded explanations and much more from opportunities for exploration and experimentation. Indeed, it is what they must do if they are to learn. Too many no's may lead not only to friction but to lasting negative notions about learning itself.

Childproofing doesn't involve just putting things away. You also need to provide alternative objects that are appropriate and challenging for toddlers to investigate. In many instances household items make the most satisfying "toys" for young explorers to fill and dump, open and close, push and pull, and fit inside one another.

Toddlers rarely play far away from their chief caregiver, so setting up a playroom or a single place for toys will only work part of the time. Rather than trying to restrict play to one room, provide a variety of play environments in various parts of the house.

Since adults need to spend so much time in the kitchen, it makes sense to stock a low cabinet—one that opens and closes easily—with toys and objects for your junior assistant:

- A few pots, lids and spoons can be used for dumping and filling, beating like a drum, or clanging like cymbals. Pots can also be worn as hats.
- Measuring cups and plastic freezer containers are interesting nesting toys for fitting and stacking.
- A few small blocks or pop beads are good for filling and dumping into pots and pans.
- Empty boxes, frozen juice cans, and paper-towel rollers are perfect for rolling, nesting, stacking, and sound effects.
- A paper shopping bag with handles is just right for filling, emptying, and carrying around for deliveries.

In all rooms of the house, move cords, lamps, ashtrays, and ornaments out of reach. Cover electric outlets with safety devices; and switch from tablecloths to placemats. Put cardboard picturebooks on the coffee table for looking at, together with a large pegboard with easy-to-grasp pieces, or a simple puzzle with knob handles.

If your bookshelves are within reach, better pack the books tight so that inquisitive little hands won't be able to budge them. Clear a bottom shelf for your toddler's own books and provide magazines you're finished with. A set of nesting blocks or a box with multiple objects to fill and empty may also hold interest for a little sit-down time.

### *Siblings: Safety and Satisfaction*

If there are older children in the family, special attention must be given to toy safety. Games, construction sets, and toys with small pieces must be stored out of your toddler's reach. Providing private storage areas for possessions gives big brothers and sisters a needed sense of privacy and protection from their sometimes pesky little relatives. This is not to say that toddlers and their bigger siblings can never play cooperatively or with the same objects. Brothers and sisters learn a great deal from each other. But a toddler is likely to handle toys with more curiosity than care, and may end up hurting himself or the older child's valued possession, or both. Some of the friction between siblings really begins to manifest itself at this stage when the toddler begins to get into things. Protecting both the older child's rights and the younger child's safety can help limit the general level of tension.

Older siblings can also be taught how to "negotiate" with a toddler. When little sister grabs someone else's favorite toy car or doll, she will usually accept a reasonable substitute. Knowing that can produce quite a different scenario than the usual tug of war and cries of "That's mine!" or "Mommy! she took my doll!" Teaching older siblings how to distract a toddler empowers them to handle toys and each other cooperatively.

## *TOYS THAT MATCH TODDLERS' NEW ABILITIES*

Few toys or activities will hold your toddler's interest for long. Variety is important. Action is the name of the game, and toys that match your toddler's new mobility are the most appropriate.

For the less-than-confident beginner, a heavily weighted walker provides something to hold on to as he walks. Galt's Walker Wagon is a good choice that will continue to be used for carting toys about. Also fun at this stage is Wonderline's heavy Snappy Snail, with chime-roller sound effects.

The sturdy Toot 'n Toddle Taxi from Wonderline is for the slightly more confident beginning walker. It's intended for the child who's fairly steady but likes the security of

*Toddler Walker Wagon*
*Galt*

holding on. In a few months, instead of pushing it he'll be straddling it and driving around the house. At that time you can remove the push bar. Doll carriage, wagons, and lightweight push toys aren't suitable at this age, as the less surefooted beginner needs a toy with heft to keep him steady. In selecting a first ride-on, avoid vehicles with wheels that stick out and get in the way of the rider's feet. Keep in mind that your toddler doesn't need an electronic ride-on that runs up the price and limits the need for foot power. A squeaker horn and feet on the floor is sportier and lends itself to active play. For a different kind of action, a low-to-the-ground rocking horse with no springs is fun to mount.

*Toot 'n Toddle Taxi*
*Wonderline*

Look for rocking horses with a limited arc that won't throw little cowpokes over. Any ride-on toy for this age group should be tested for stability and tippability. Your toddler's feet should be able to touch flat on the ground when she is seated. Wheels should be spaced wide apart, and four wheels are less likely to tip than three. Toddlers don't need pedals or a steering mechanism. But they do enjoy features like storage compartments so they can carry along small toys; and they love ride-ons with sound effects like horns, or ratchets on wheels. They also love ponies that jingle and trains that toot.

Both boys and girls enjoy toy cars, trucks, and planes that they can roll along the floor. But stay clear of Hot Wheels and Matchbox miniatures for now—their small parts can be hazardous. And oversized metal trucks are still too heavy and complicated for this age group. Keep the vehicles simple. Little Tikes' Toddle Tots Family Car, with Mom, Dad, Brother, and Sister figures to load and unload, is a good choice.

## *Push before Pull*

As toddlers become more surefooted, push-along toys with a stiff rod make taking a walk into a game. A classic Melody

Push Chime or colorful Corn Popper, both from Fisher-Price, offer just enough sound and motion for beginners who still like the security of holding on to something and delight in making things happen.

Because looking back over the shoulder is difficult for inexperienced walkers, pull toys are for the slightly older toddler who really has her land legs. Again, motion and sound effects have great appeal. Although some of the most handsome pull toys are wooden works of art, they are also terribly costly for the short-term use and limited novelty they provide. Johnson and Johnson's Shape Sorter Transporter is part puzzle and part pull toy. So it is more versatile and provides multiple ways of playing. The toddler can use it both for action and sit-down play with shapes to stack, fit, and deliver.

For safety's sake, be sure that pull toys have no bead handles that can be mouthed and no more than a ten to twelve inch cord for pulling.

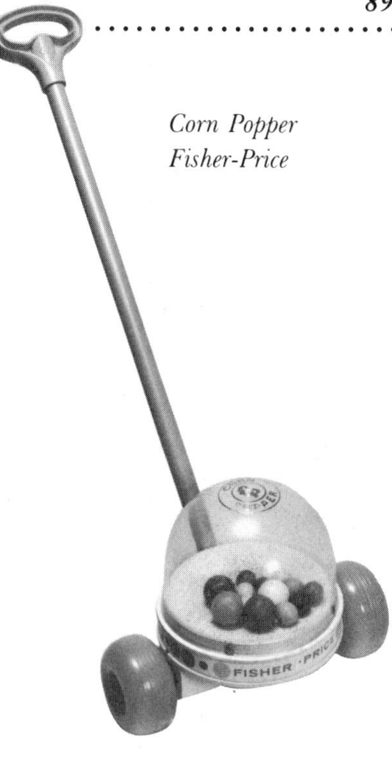


*Corn Popper*
*Fisher-Price*

*Shape Sorter Transporter*
*Johnson and Johnson*

## *Balls*

Balls of various kinds go right along with your toddler's active play style. Big beach balls and easily grasped cloth or vinyl balls are great for games of rolling the ball back and forth. Playing ball with your toddler also introduces him to the principle of give and take. Indeed, it's the first cooperative game children learn.

Since your toddler's aim is anything but accurate and her balance is tentative, stick to lightweight balls that won't bop her (or nearby objects) off balance. Foam rubber balls may be dangerous if your toddler is still chewing toys. The fabric balls recommended for infants will still be of interest.

Ping-Pong balls are easy to grasp but also easy to bite, and break when stepped on. Since most toddlers still mouth toys, they are better avoided.

## *Climbing, Sliding, and Swinging*

Because there are few toys more fascinating to active toddlers than steps, families who live in split-level or two-story houses really need to protect young toddlers from dangerous falls. Some experts advocate no gates but rather suggest parents teach babies early how to climb up and down. However, others feel gates should be provided to protect children both at the top and bottom of stairways. Unfortunately many families still have old-fashioned gates that can trap children's heads. Don't use gates that have been stored in the attic. Buy new ones that meet the latest safety standards. New gates have straight top edges rather than the old-fashioned zig-zag tops of the expandable design. Putting safety gates up leaves your child free to play without the need for you to hover at all times. This doesn't mean a toddler should not learn how to climb up and down. It only means he shouldn't do so without supervision.

Of course, stairs are not the only things toddlers climb. Tables, chairs, couches, windowsills, counters, and even bookcases are all fair game if your toddler can get a toehold. Tippable stepstools should be removed and unstable chairs and tables packed away. But declaring certain areas off limits to climbing doesn't mean all climbing should be stopped. You probably couldn't do that even if you wanted to. Rather than making climbing into a forbidden game, provide your toddler with plenty of practice time. For a toddler eighteen months and older you may prefer a low indoor climbing gym. The traditional Childcraft wooden Toddler Gym is designed for young children to climb, slide down, and hide in. Be careful, though. Many of the older versions of this type of play gym present a strangulation hazard. If you find an old one in the attic or at a garage sale, it may not meet today's safety standards.

***Caution: If you have a play gym made by Creative Playthings before 1980, don't use it! Nor should a similar model that was sold by Sears Catalogue be used until a safety shield and new ladder are installed. Sears will provide these free of charge if you own the old model.***

*Toddler Gym*
*Childcraft*

*Toddler Swing*
*Little Tikes*

You will need to be on hand while your toddler is learning to use the ladder and slide. This piece of equipment may be too bulky for a small home or apartment, but many parents who live in homes without stairs feel that it provides a playful way of gaining motor skills while having fun. Don't feel you must buy such toys, however. Children who don't master climbing steps now will do so a little later on, either at the playground or in other people's homes.

Children under two also enjoy the thrill of moving back and forth through space on a swing, although it may be a bit soon for a full-scale swing, slide, and see-saw set in the backyard. For safety's sake, look for a well-made toddler swing with a molded seat and a safety harness inside. If you have no backyard, go down to the park. There will be plenty of swings and lots of other children.

## *Dolls and Stuffed Animals*

Both boys and girls enjoy a doll to lug and hug even if they don't have a carriage. Soft, uncomplicated dolls are easiest to hold and play with. Clothes are not important since they are too difficult for toddlers to manipulate. A box or basket can be made into a cradle for putting the "baby" to sleep. Remember that most toddlers become especially fond of one particular doll or animal, and that this attachment frequently continues for a long time. So it's a good idea to select a doll that can take repeated cleaning without falling apart. Of course, there's no way of predicting what stuffed animal or doll your child will form an attachment to. As a parent, the best you can do is provide choices that have no hidden prickles.

As suggested in the previous chapter, soft dolls need to be examined carefully in terms of safety. Many toddlers are still apt to chew button eyes, bells, and other doodads. They may fall asleep on top of a doll or roll over on it. Dolls with

OPPOSITE
*Puffalumps*
*Fisher-Price*

small parts—zippers, whiskers, or wire under the plush—are still potential hazards. Look instead for well-made dolls that are safe enough to leave with a child when you leave the room.

Fisher-Price's Puffalumps are lightweight but have the heft appeal toddlers love. The new breed of Puffalumps with glasses are not recommended, however.

## *HANDY TOYS: MANIPULATIVES*

*Rings and Rollers*
*Johnson and Johnson*

For infants, simply holding, twisting, or tasting a toy offers enough manipulative play value, but toddlers like their manipulations to lead to some kind of outcome. In other words, a toy should not just have an action but a reaction, too. At this age, your toddler has considerably more eye-hand coordination than he did a few short months before. Nevertheless, toys need to be chosen that challenge without frustrating. It's easy for him to pull pop beads apart but still difficult to put them together. Most toys with interlocking pieces will be better for later in the year. A set of nesting cups or ring-and-post toys can be used as multiple objects to roll, chew, bang, toss, or carry about. But fitting them together according to size will be beyond your toddler's abilities for now. The problem with buying your child such toys too soon is that they may hold no interest by the time they are most appropriate.

One solution is to present only a few pieces of the nesting cups or only the rings of a post toy for now, and then add the rest later. Or consider Johnson and Johnson's Rings and Rollers, a ring toy with rings of only one size, so there is no order that's right or wrong. Chances are the marble-filled rollers will attract more attention than the rings, but the fitting principle is there when the time is right.

At this point, activity boxes that used to be attached to the crib rail often become favorite floor toys. Given a toddler's low tolerance for frustration, it may be that these so-called "baby toys" offer just the right ease and sense of mastery she most enjoys.

For the most part, young toddlers are more interested in motion and big muscle action than in prolonged play periods with toys that demand fine motor control. They like both small and oversized objects to carry about with them. Clutching two small dolls—one for each hand—seems to give some young walkers another way of "holding on." Soft clutchables are a safer bet than hard objects that might hurt them when they trip.

*Jack-in-the-Box*
*Galt*

## *Hinged Toys*

Most mechanical toys demand more dexterity than your toddler has. A jack-in-the-box with a crank is beyond him, but a closed box that opens with the push of a button offers both a challenge and a gentle surprise.

Playskool's Busy Playhouse combines action and sound with lids to lift, a spinner to twirl, a bell to ring, and Kiddi Links to drop down the chimney.

*Busy House*
*Playskool*

Hinged toys with doors that open and close lend themselves to manageable manipulation. Don't overlook the possibilities in household items that slide, lift, and open in different ways. Band-aid boxes, empty cartons of cotton swabs, and see-through boxes make interesting puzzles for little fingers. Put a cracker or raisins inside for an edible treat.

## *Shape Sorters*

Judging from the number of shape sorters available, one would think that sorting shapes is the dominant interest of toddlers. There are ride-on toys with shape sorter cut-outs in the seat as well as turtles, hippos, trucks, cubes, mailboxes, pull toys, and bath toys with the same shape-sorting feature. While this extra feature adds another dimension to a plaything, shape sorters should not dominate their playthings. In fact, many of them are too difficult for the very young toddler.

By the middle of their second year, toddlers do enjoy fitting objects into other objects—the hide-and-seek aspect of putting toys into containers and emptying them. But they don't need dozens of toys to hone such skills. Homemade shape sorters can be made easily with cut-outs on a shoe box or the plastic lid of a coffee can.

Often the store-bought variety features far too many complex shapes for the young child to handle. A shape sorter with no more than three shapes offers enough challenge for beginners. Save the more complex sorters for later. Battat's Sound Puzzle Box gives a whistling payoff when shapes slide down the see-through chutes; and it can also be used for a rolling toy.

Or consider Playskool's My Shape 'n Stir Pot that will be used for filling, dumping, shape sorting, and early games of pretend.

*My Shape 'n Stir Pot*
*Playskool*

Knowing the names of abstract shapes isn't particularly important to children's early experience of language. Use the words, but don't push them. Knowing the names of shapes at two is not a ticket to Harvard or Yale—it simply means a child has been drilled over and over again, often at the expense of more active and interesting conversation and play. Far more useful is exposing children to plenty of sensory experiences with balls, apples, blocks, containers, boxes, and wheels. That way they learn multiple meanings of the words "round" and "square." Pressing for words before meaning is like teaching a parrot to count to ten. It can be done, but what's the point?

## *Blocks and Construction Toys*

A set of large nesting blocks of sturdy cardboard or bright plastic can be used in many different ways both now and later. Probably the last thing your toddler will do with them is to nest them in the proper order, but that feature does make them compact to store. Besides, once you've put them away, your child will undoubtedly enjoy removing them! Just as toddlers learn to undress before they learn to dress themselves, some toys can be enjoyed in reverse order. Similarly, your child will enjoy knocking a stack of blocks over long before she can stack them back up. If you haven't the time or patience—nor an older child to do the building—save the blocks for later.

A set of nesting blocks is perfect for hide-and-seek games with objects. Cover a toy ball or small bear with the big block and ask, "Where is it?" Once your child has got the hang of it, move the blocks around and see if she can still find the missing bear. Of course, you don't need to buy nested blocks. If you prefer, you can collect assorted boxes or cans and cover them with adhesive paper for decorations. The store-bought variety may be sturdier, but the home-made kind can easily be replaced and will probably please your child equally.

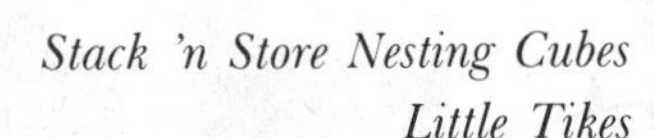

*Stack 'n Store Nesting Cubes*
*Little Tikes*

A box or wagon of small wooden blocks is fun for filling and dumping. Cube-shaped blocks are easy for young toddlers to grasp and eventually stack. The numbers and letters should be regarded simply as decorative features. By the time children are ready for letters and numbers they'll have outgrown their interest in these blocks.

Older toddlers will also enjoy the repetition and motion of a hammer board, which combines action with eye-hand coordination. For safety's sake, buy a pounding board with pegs that slide but don't come out. Little Tikes' Flip Flop Tool Box is easy to tote. After the pegs are pounded down, the board flips over and the pegs can be pounded again. For long winter days indoors, when cabin fever sets in, hammer boards are an excellent outlet for energy.

Lugging large objects from one place to another is an appealing motor activity. This is a good time for a set of

*Super Building Blocks*
*Kettering*

lightweight cardboard blocks. Super Building Blocks from Kettering take time to put together but are sturdy enough to last through years of play. For now they have enough heft to satisfy young haulers. Just a few make a mighty tower that won't hurt when it gets knocked down. They're strong enough to stand or sit on, and they'll combine well with wooden block constructions a few years from now.

By eighteen months your toddler may find interlocking construction toys, such as Lego's Duplo blocks, challenging. Again, in the initial stage, the snap-lock feature of these plastic blocks is likely to lead to more taking apart than

*Duplo Blocks*
*Lego*

putting together. If you have older children who are busy with Lego-type construction toys, your toddler may be eager to have a related toy. But be careful: the regular-size Legos present a real choking hazard to toddlers. Happily, though, an investment in oversized Duplo pieces has long-term value since they combine later with regular-size Legos. The same is true of Tyco's toddler- and standard-size construction blocks.

## *Tub and Water Toys*

A few playthings in the tub can enhance the enjoyment of water play. Play Buckets from Kiddicraft are easy to grasp. They come in a set of three: one has a spout, another has a hole where the mouth should be, and the third has a sieve bottom. These can be used in the tub, and later, in the sandbox. Toddlers will also enjoy the Fisher-Price Blue Bird with multiple play pieces to use for floating, spilling, filling, and squirting.

OPPOSITE
*Play Buckets*
*Kiddicraft/International Playthings*

Water, by its very nature, is fascinating to splash, pour, sprinkle, and taste. With an assortment of toys, toddlers can make a duck swim, propel a boat, or bathe a doll. Many household items are also ideal for exploring the nature of water. Measuring cups are fun for pouring or floating, a sponge is good for squeezing, and empty plastic bottles are great for squirting. A plastic funnel and sieve are fascinating for experiments in filling and spilling.

Of course, you won't want all these things in the tub at one time or crowding the edge of the tub in between baths. Change the supply from time to time. You can hang them out of the way in a net shopping bag.

There are a great many activity boards that attach to the tub wall and operate like crib-side activity boxes. Oddly enough, none of the parents I interviewed mentioned them among their child's favorites. Frankly, for the tub I prefer an assortment of objects in the water to a fixed plaything children need to stand or stretch to reach. While such toys may be interesting when new, they are more expensive and less versatile than floating and sinking water toys that can be used in multiple ways.

Fortunately, water play need not be limited to bathtime. Toddlers seem to have radar that searches out puddles to stamp their feet in. The small tub you used to bathe your child when she was an infant is good for washing toy dishes, sailing boats, and floating sponges. Shovel and pail, of course, are basic equipment for a day at the beach. More elaborate sand and water toys will add enjoyment next year. For now keep the objects simple. Sand and water are themselves new and fascinating enough to provide hours of delight.

## *Books as Toys and More*

Perhaps the best "hinged" toy of all is a cardboard picture book with an interesting illustration on every page. By now you've probably started reading aloud to your child. Toddlers enjoy both listening to simple stories and looking at illustrations.

Long before they speak very many words, young children understand and enjoy little books that show familiar objects and events. They especially like the rhythm of nursery rhymes. Before long they'll enjoy playing *I Spy* games, finding the things you name on the printed page. Also interesting for sharing are books with flaps that lift up to serve as a peekaboo game, as well as books with textured pages, like the classic *Pat the Bunny.*

Some of the books you share may be too easily torn for independent "reading." Be sure to provide some sturdy books that can be picked up, carried around, and explored solo.

By midyear, toddlers have usually gotten a firm foothold on mobility. Once they reach this developmental landmark, their use of language begins to blossom.

By the end of the second year, many children have as many as two hundred words in their vocabularly. In another year they'll be using more words than you can count. While manufacturers of "educational" toys often tout the importance of particular toys for language development, it's not the toys but *how they're used* that enhances a child's learning. A toddler with a roomful of toys but no one to talk with will not learn more than a child with relatively few toys who is surrounded by people who interact, talk, read, and respond to him.

## *Music, Singing, and Dancing*

From their earliest days, infants enjoy the rhythm, motion, and sounds of music. Toddlers who have been sung to will be singing along toward the end of their second year. Keep the songs simple with plenty of repetition.

You can also put on a recording and dance together. Ring-around-a-rosy puts a little humor, music, and action into the inevitable business of falling down. Toddlers love to dance, conduct, and bang along with a spoon and pot lid. They also love the magic and power of helping to turn the music on and off. Although he can't operate it independently yet, this may be the time to buy him a tough tape recorder. The Fisher-Price Tape Recorder is simple and rugged enough for preschoolers to use solo and for toddlers to enjoy with a little help.

Also fun for your toddler's musical explorations is Battat's Baby Rhythm Band. This colorful, one-piece musical activity center includes a xylophone, bells, and drum. Toddlers will love their noisy experiments in cause and effect with plenty of sensory feedback.

## *Early Pretend-Play*

During the latter half of the second year, children begin their earliest games of pretend. At this stage, making believe is more imitative than inventive. Toddlers begin by doing what they see adults doing.

Your child will enjoy a toy telephone almost as much as a real one. Although there are expensive toy phones around that feature TV characters who talk, for the toddler a simple phone that's light enough to carry is preferable. At any age, it is certainly more versatile. The talking phones may have novelty appeal, but both the novelty and the batteries wear out. And for the older child, the preset conversation may actually limit the kinds of dialogue (and monologue) he is perfectly capable of creating. At any age, the problem with talking phones is that they do the talking while the child does the listening. What you end up with is a toy that costs more and has less play value.

*My First Phone*
*Ambi*

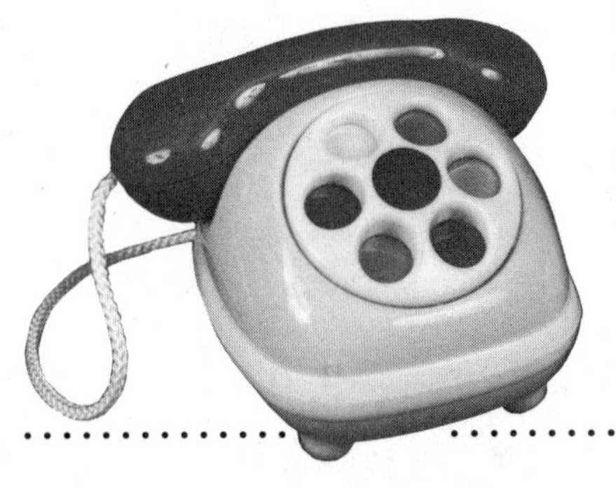

*Toy Dishes*
*Bambola*

Since few toys are more desired than Mommy's or Daddy's keys, a ring of plastic keys may provide an acceptable substitute. Most of the activity toys that involve fitting keys into keyholes are too difficult for young toddlers to handle. A simple ring of plastic keys won't fit anywhere, but they're not bad for chomping on, turning, and carrying, and they're a lot safer than real metal ones that can jab or choke.

Another push-along toy with plenty of potential for pretend-play is a small shopping cart or doll carriage. There are beautiful low wooden carriages and charming wicker imports. Some tend to be pricey, however, and are quickly outgrown. Little Tikes' pink plastic Doll Buggy is a practical choice. Older toddlers love to take plush animals and dolls in and out of their buggies. And a small blanket to put on and off is about all a well-dressed—and undressed—doll needs.

Though full-scale tea parties and elaborate doll play will be increasingly interesting next year, toddlers enjoy playing with the multiple pieces of a study set of dishes and pots and pans. Avoid anything but stout plastic dishes that can be washed, stepped on, and otherwise abused. They don't need toy spoons, forks, or small lids that may be mouthed. Remove these from any set.

*Table and Chairs*
*Little Tikes*

Toward the end of their second year, as children become fussy about sitting in high chairs and eating in general, a low table and small chairs will be useful equipment. They can

be used for real meals as well as play. It's a great place for doing puzzles and using art materials, and for tea parties, too. Since one of your child's favorite games will be to wash the table top, look for a surface that can take frequent sudsing as well as finger paints, clay, and other messy substances. Although sitting is not an activity toddlers engage in for long stretches, a table and chairs is a good investment for years of play value.

## *Early Art Exploration*

In the latter part of the year some toddlers are ready for simple art materials. As with most other playthings, manipulation and exploration are what motivate them, not the desire for "artistic expression."

Children who are still chewing playthings may not be ready for crayons, markers, or chalk. And since it takes time for children to learn where they should and shouldn't color, art supplies are best used with adult supervision. If your child is not ready to accept the ground rules and repeatedly tastes the crayons or scribbles on the walls or furniture, put the crayons away for a few weeks or months.

Scribbling is what toddlers should do with crayons, markers, pencils and pens. They don't need lessons now in how to grasp a "drawing tool" properly. Nor are they ready to write letters or color inside of lines. Blank sheets of paper, big pieces of wrapping paper, and big sheets of newsprint are fine for these early adventures in art. Choose jumbo-size crayons for ease of handling. Binney and Smith's So Big Crayons are the right size for now. Slender, standard-size crayons are likely to snap in clenched little fists and fingers. Ideally, crayons for toddlers should have one smooth flat side to prevent them from rolling. Some school supply catalogues have them. More recently some stubby crayons shaped like single-headed dumbbells have become available. They're easy to grasp and won't roll too far, but

*Crayola So Big Crayons*
*Binney and Smith*

they're expensive and not widely available. Whichever kind you buy, be sure that they're marked "nontoxic," and for safekeeping put them away when they're not in supervised use.

To pass time in the doctor's waiting room, a notepad or the back of an envelope and a pencil or pen make a welcome diversion. Since toddlers especially love using "real" tools they have seen grown-ups use, pencils and pens have special appeal. However, the sharp points can be hazardous, so their use needs to be supervised. You may also want to make felt-tipped markers available. Since they produce bright and fluid drawings, children love them—but, again, they are not for independent play. In addition to the fact that the caps present a choking hazard, a marker left uncappd will dry out and become useless. Besides, young children are apt to get carried away, marking up walls, objects, and themselves. Even watercolor markers can leave stubborn stains. Introduce them if you like but be prepared to be there or accept the consequences.

At eighteen months, children enjoy pounding, pinching, pulling, and poking at Play-Doh or a homemade variation. You'll find the recipe in chapter 4. They'll be much less interested in making something than in exploring the way this plastic material responds in their hands. Of course, keeping Play-Doh in one place, so that it doesn't end up in the rug or on the furniture, can be a problem with toddlers. And for children who still chew everything, it may be better to postpone the introduction for a while. The fact that products are marked "nontoxic" doesn't make them desirable eating.

Least expensive of all and virtually trouble free, "painting" with water is both safe and satisfying for outdoor fun. A big paintbrush and a bucket of water lend themselves to the kind of repetition toddlers enjoy. When they paint the fence, the house siding, the sidewalk, or a wagon, the wet "paint" changes the color of surfaces enough so that children feel they've made something happen. The fact that it's all gone a few minutes later doesn't matter so much at this age. It's the doing, not the end product, that matters to young children.

## *SUMMING UP*

As toddlers approach their second birthday, their repertoire of play is growing rapidly. They no longer wait for someone to initiate a game or offer a plaything. With both language and locomotion they are able to engage themselves and others in play. Increasingly they have preferences and a style of playing that reflects their individuality. Their favorite word—"No!"—comes with their growing sense of independence—a small but powerful declaration of their newly discovered status as a separate person. No child needs all the toys suggested in this chapter. However, the right toys and the freedom to play can enhance their sense of independence and their enjoyment of interdependence. Socially, intellectually, and physically they have come a long distance from that largely dependent infant who needed to be carried from one place to another and whose cries needed to be interpreted and often guessed at. With the power of words and the ability to express what they want and need, the older toddler is on the verge of becoming a less impulsive, more reasonable, and even more playful little person.

# IV. *Toys for Older Toddlers: Two to Three Years*

## *NEW BOUNDARIES*

By two, walking is old hat. Now she runs, jumps, climbs, and slides. Though more surefooted than before, she is still something of a klutz. She doesn't always look before making a mad dash, and ends up tripping on the toy she has just discarded. As she climbs to the top of the slide, she makes it clear she wants no help or interference; a few minutes later she needs help getting down. Since she is clumsy, a supply of band-aids is needed to mend the boo-boo's and soothe her wounded pride.

This tug of war between the desire for independence and the continuing need for dependence is what gives older twos a reputation for being "terrible." In fact, the twos are more "testy" than terrible. What they're testing are their own limitations and yours. They neither want nor can they handle the total independence they keep asserting.

Bursting with a new sense of himself as a separate person, your big toddler is easily frustrated when his skills fall short of the mark. He can get most of his clothing off, but getting it on is another story. He needs help moving his toy car, but when you try to help, he says, "Me do it!"

As his language blossoms it's much easier for him to express what he wants and how he feels. As he increasingly makes his wishes known, some of his negativism will be reduced. He still says "no" more often than not, but he can also say "yes" or "okay." He talks in short sentences not only to you and others, he talks to himself, especially when he's playing alone. Listen in and what you'll hear is an abbreviated play-by-play commentary: he is telling himself what to do.

At play, he provides inanimate objects with lively sound effects. He does not yet fully understand that an electronic teddy bear that talks and moves is not alive. He believes that objects, especially those that can move, can also see, feel, and think just as he does. In fact, he believes everybody thinks, feels, and sees what he does. Being the center of the universe, he cannot imagine any other point of view but his own. He still has so much to learn about what's real and not real, what's possible and impossible, what's right and wrong.

## *New Use of Language*

In trying to pin down the big and sometimes confusing world of things and people, language is used not just to communicate but to support thinking and putting ideas together. If you live near a farm, your child knows that the big creatures in the field are cows. Take him to the zoo and he will call the first bear he sees "cow!" His lack of experience leads him to overgeneralize. If Mommy usually carries crackers in her pocketbook, he expects crackers to always be in pocketbooks. When Mommy opens her bag and says, "No crackers here today," he'll go to Grandma's pocketbook expecting to find crackers.

One can almost hear and see the wheels of thought turning as twos go about. Out of sight is no longer out of mind. If he was playing ball before going to the store with Daddy, he can remember when he comes home where the ball is and go get it. When there are no crackers in Mommy's or Grandma's pocketbook, he thinks for a moment and says, "Cookies in kitchen." His memory is growing and so is his ability to make associations.

Twos don't just climb stairs or open and close doors because they're there. They go upstairs or open a drawer to get something they want. Nor do they use words simply to label objects. At this age, words can also be used to bring images to mind and to complement thought. Words give the toddler a new kind of power. He not only knows he wants a drink, he can say he wants "juice" and show you where it is. Though his vocabulary is still limited, a single word or phrase can be used for multiple meanings.

In his efforts to make sense of this big and sometimes confusing world, the toddler is still struggling. He's intolerant of many changes. He loves rituals because of their predictability. They provide order and a sense of control he wants but still lacks. "Mine" is not just a favorite word, it's an expression of his growing sense of selfhood. The inability of young twos to share is not so much a sign of selfishness as it is the child's own way of defining himself. Knowing what's "mine" definitely comes before any consideration of "yours." Very young children are very rigid in their thinking. Mommy's keys, Daddy's hat, and Johnny's pail are tightly defined, and any deviation from that orderliness is disturbing.

If you're reading one of his favorite stories, he'll know if you change a word or skip a page. At bedtime he may have a nighttime ritual that cannot be altered without setting off wails of protest. The fact that he can remember what comes next or who goes with what is a good indication that his memory is developing, and along with it a fuller sense of selfhood.

### *New Uses of Toys*

By two and a half, toddlers begin to use familiar toys in original ways. She puts the doughnut from her ring-and-post toy on her wrist and calls it a bracelet. She eats a cookie and offers a bite to her teddy. They are still very literal, and more imaginative flights of fancy are yet to come. Even so, the ability to use one thing to stand for another marks the beginning of symbolic play; it's another signal of a new and more advanced level of intellectual development.

Twos are a study in contrasts. They are no longer the dependent babies they once were nor the independent children they will become, but rather a bit of both. This is an age of transition when the skills that began to emerge earlier are refined and expanded. Though few toys will hold their interest for prolonged periods of time, they are ready for new kinds of playthings to match their growing physical, intellectual, and social skills.

## *TOYS FOR PHYSICAL DEVELOPMENT*

Through their active play, children develop a sense of their own physical competence. They are keenly interested in testing their capacities for climbing, running, and walking under, over, and through a variety of spaces. This active kind of play not only helps them develop their gross motor coordination, it strengthens their new sense of selfhood and proves a good outlet for their seemingly endless energy.

Although some toddlers are more active than others, almost all twos relish time and space to shout and run and use their big muscles. A romp out of doors is a definite plus, especially if there's sand or snow to dig, wheel toys or swings to ride, a low slide or gym to climb. Remember, though, that twos are still limited in judgment. They may get to the top of the slide and not know how to get down. They don't watch out for swings or equipment that could bop them. If a ball rolls into the road they'll run after it. As a result, they need to be closely supervised, both in the backyard and at the playground.

*Tike Treehouse*
*Little Tikes*

## *Climbing, Balancing, and Swinging*

A low climbing device with a few steps up to a platform to stand on and a small slide down is a good choice for indoors or out. The climbers suggested earlier will still fit, but if you waited, you may prefer the slightly larger Tike Treehouse from Little Tikes. It will be used for social and dramatic play as well as for physical exercise.

Twos love to balance themselves on curbstones and low garden walls like a tightrope walker. A walking plank made from a simple but smooth two-by-eight can be used indoors or out. When walking forward gets too easy suggest walking sideways or backwards. For a further challenge, raise the board with two fat phone books.

Few things are as satisfying to toddlers as the soothing motion of a swing. For safety's sake use a toddler chair swing with a back, sides, and safety strap. Your two-year-old can't pump but he'll love to give orders for "more" and "again."

*Chugga Chugga Choo Choo*
*Little Tikes*

## *Wheel Toys*

Most of the ride-on toys suggested for the younger toddler will still be appropriate now, but steer clear of oversized kiddie cars that are hard for this age to propel. Four-wheelers with good balance are still the way to go. Little Tikes' Chugga Chugga Choo-Choo with an engine and a little red caboose has sound effects in the wheels and plenty of storage space for freight. And the wheels are inset so that they're not in the way of the child's foot action. That's not true of all ride-ons. That's why it's best to test-drive before you buy riding toys.

Most toddlers will be quite content on four wheels, and safer, too. Rather than rushing into three-wheelers, consider some of the other wheeled toys that fuel imagination and have multiple uses. For action-packed play, a low wheelbarrow or Mattel's mini-shopping cart are great for pretend deliveries. Or, for useful pull-along hauling, provide a wagon that's lightweight and small enough for a young child to handle. A classic red Radio Flyer in the smallest size will fit, or for a lighter version with a pull-string handle, look at Little Tikes' toddler-size wagon.

By two and a half, some children are ready for their first pedal toy. The first trike is likely to be outgrown quickly, but for safety purposes it should be small and low to the ground. A ten-inch front wheel is appropriate for most two-year-olds. When testing a trike, remember that your toddler should be able to reach the pedals and get on and off easily. Most kids take their first ride on a trike at a friend's house, at the playground, or in the toy-store. Take your cue from the way she handles those test drives. If she seems to have some notion of how the pedals work—and the coordination to make them go—you'll know she's ready.

There's no point in buying something to grow into if it's more frustrating than fun. A trike that's too big may also be a real tipping hazard, so don't be penny-wise and pound-foolish.

## *Balls*

Balls of many sizes and shapes are fine for indoor and outdoor fun. Twos can really throw now, and if the ball is light and big enough they may even catch a well-placed throw.

Beachballs, foam rubber balls shaped like footballs, basketballs, and soccer balls are good for chasing after, kicking, throwing, and even for aiming into empty boxes or knocking down blocks.

## *Boxes and Other Interesting Spaces*

One of the most versatile toys for indoors and out is a big cardboard packing box. It's fun for climbing in and out of, for tunneling, for filling and dumping, and even for hauling. You don't need to do any decorating to "sell" it. Just the sheer size and an open end is inviting. Some beverage boxes come with cut-out hand grips toddlers will enjoy. Tie a rope to one side and you've got a wagon that will slide on bare floors. Turn it over and it's a great little table for drawing on. Boxes like these make a wonderful refuge or secret place to hide away safely. Just be sure that staples and sharp edges are removed and you're in business.

*Tunnel of Fun*
*Childcraft*

Good fun for crawling through or hiding in is the fabric-covered Tunnel of Fun from Childcraft. If you've got an old one from the attic, check the fabric label. The new models use flame-retardant fabric. Old ones may be hazardous.

## *Music and Movement*

By two, most children have a repertoire of favorite songs and enjoy chiming in with a word or a phrase. They not only sing, but they love to dance, gallop, clap, tap, and turn till they're dizzy. They like singing along with Mommy or Daddy. Stick to lyrics that are repetitive and that have lots of cues for joining in. Songs that call for action that's easily mimicked can become a good game. "Put Your Finger in the Air" is a playful example that twos can follow. Tapes and records of nursery and folk songs are fun for singing along, marching and dancing.

*Tap-A-Tune*
*Little Tikes*

Also nice, if you can handle the decible level, are percussion instruments such as drums, tamborines, or simple rhythm sticks. Homemade instruments can be improvised without much effort. Empty containers taped shut with some cereal inside make excellent maraca-like shakers. You can, of course, buy rhythm instruments. Most toy-shop versions have poor sound qualities. You'll find cymbals to crash, a drum to beat, and maracas to shake in school catalogues and music shops.

Twos will also enjoy making music on the sturdy Tap-A-Tune Piano from Little Tikes. It has a better tone than most. Save the color-coded music for future use.

Introduce your children to a variety of musical genres with records and tapes. Let them dance, conduct, or clap along. The music doesn't have to be "for children." If Sesame Street is on their hit parade, bring some home by all means. But don't forget folk songs and marches. Add some symphonic music, a touch of jazz, or rock and blues. Music puts no age limits on enjoyment.

## *Sand and Water*

Indoors and out there are few play materials better loved by twos than sand and water. Rich with sensory experiences, both change, move, and respond to the touch. Such hands-on experiments provide the meaningful underpinning for expressive language. Abstract words like wet, rough, drippy, smooth, and soft become meaningful when coupled with real experience. With the addition of "tools," both sand and water can be poured, patted, pounded, poked, and otherwise enjoyed.

For the sandbox, a pail and shovel are basics. Twos are also ready for more elaborate experiments. A sand mill, a sieve, strainers, and plastic dumptrucks and earthmovers will make roadways and hills. Plastic containers of various shapes and sizes from the kitchen also make handy sand toys. Molds are a bit tricky for twos. Save them for next year.

*Sandmill, pail, and shovel*
*Bambola/International Playthings*

Water play is so satisfying and irresistible that few twos can pass a puddle without stamping a foot in it, or floating a twig or sinking a pebble.

For the bathtub or outdoor wading pool Johnson and Johnson's Bath Time Water Works has multiple play pieces for experiments in squirting, pouring, and sprinkling. Small boats with dolls and animals that fit inside are fun for launching games of pretend. Young twos can't operate wind-up toys just yet. Their own actions supply enough reaction to satisfy. Many of the bath toys suggested in the last chapter continue to be of interest.

Keep in mind that water play doesn't need to be limited to the tub or beach. A toy sink or plastic dishpan placed on a towel-covered table is ideal for whipping up a storm of suds and washing dolls' clothes and dishes. True, it's messy but a towel on the floor will blot up the drips while the water provides a medium for sensory learning, dramatic play, expressive language, and just plain fun.

Water play, of course, should always be supervised whether children are *in* water or *near* it. Under no circumstances should they be left alone, even for a moment.

## *Art Materials*

Two-year-olds enjoy experimenting with crayons, markers, and paper. By midyear many are ready for tempera paint with big brushes, or the more direct approach of finger paints. Be advised, though, that using paint with twos can be messy. Some toddlers simply aren't ready to live with the restrictions that painting requires. If they can't enjoy the paint without walking around and dripping a trail, better to pack the paints away and try again in a few months.

You can make your own finger paints with the recipe below or buy a set that comes complete with specially treated paper. Since the sensory act of finger painting is more important than the finished product, don't worry about getting it onto paper. A plastic-topped table makes a wonderful "canvas," large enough for the broad hand and arm movements that squishy finger paints provide.

Instead of a tabletop you can also use Childcraft's Messy Play and Hobby Tray that provides a self-contained surface with raised edges. It's easy to clean and can double for other art explorations with clay, crayons, and markers. If saving a "masterpiece" is essential, children can work on paper or you can get a print by placing a sheet of paper on the tabletop masterpiece and then lifting off. This is a tad tricky —but it introduces the idea of printmaking to the play.

***Recipe for Finger Paints***

3 tablespoons cornstarch.
3 tablespoons cold water.
Mix to make a paste.
Stir in 1 cup boiling water.
Stir until smooth.
Add vegetable coloring or nontoxic tempera paint to desired color.
Cool and play.

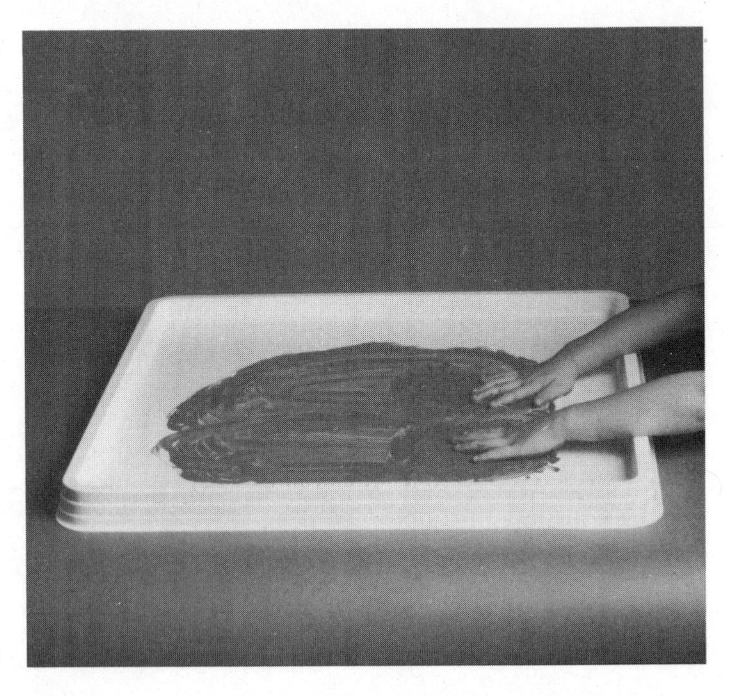

*Messy Play and Hobby Tray*
*Childcraft*

Whether they're working with crayons, markers, paint, or play dough, the big excitement is in the doing, not the finished product. Save some of these early masterpieces for yourself if you like, but products don't mean much to a two-year-old. Keep in mind, too, that some twos really hate getting their hands dirty with finger paints. There's no reason to push it. Play dough or paint and brush may be more comfortable mediums for them, and they are less fuss and bother for you.

Rather than buying an elaborate easel, put the paints out on a low table with a scrubbable surface. Children have better control of the paints and fewer drips to contend with on a flat surface. If the table is child-size they can stand up and use their whole arms for broad strokes. Big sheets of newsprint are relatively inexpensive and available at any art store. In the interest of mess prevention, art material should be an adult-supervised activity. A mat under their feet and spare paper under their art will blot up inevitable spills and drips.

Various art supply companies have come up with paint containers and brushes to minimize messiness. Adica Pongo packages Squish Paint in pans that are like old-fashioned inkwells. There are also paintbrushes that contain a supply of paint in the handle. Certainly such items are more convenient—and this may cause some parents to be less reluctant about making paints available regularly—but they are more expensive and provide a different *kind* of painting experience. It's difficult to mix colors or get a variety of strokes when the materials are self-contained. Eventually children should have opportunities to experiment with free-flowing paints and big brushes. Forget about watercolor or tempera solid paint cakes for now, though. They don't produce vivid colors and are hard to use for twos.

Play dough, homemade or otherwise, continues to be a desirable choice for three-dimensional art explorations.

*Recipe for Play Dough*

1 cup flour.
1/3 cup salt.
A few drops of vegetable oil.
Enough water to form dough.
Optional: dough can be colored with a splash of nontoxic tempera paint or vegetable coloring.

Some older toddlers may also enjoy the slicker texture of Plasticine, a nonhardening modeling clay that comes in bright colors. Unfortunately, Plasticine is a bit hard to manipulate until it warms up in your hands, so you'll have to get it started for them. As with any art material, you'll want to be sure it's nontoxic. However, even if it is "safe," you won't want the child ingesting much of it. If your child persists in tasting, you'd probably do well to put the material away for a few months. Or you may prefer to mix an edible dough:

1 cup smooth peanut butter.
1 cup powdered milk.
1/2 cup honey.

Although older toddlers may begin naming their creations, their efforts tend to be more accidental than purposeful. The child doesn't start out with the idea of painting a zebra. It's not until the stripes are on the paper that she says "Look! A Zebra!" Though some children will do this kind of labeling, the end product remains less important than the delight of experimenting with color, shape, and movement. In other words, doing is more important to twos than preserving their creative explorations. As a result, a chalkboard, chalk, and sponge make a wonderful wipe-off-and-do-again medium. There are chalkboard carry cases, table tops, and hang-up boards. The bigger the surface, the better for big arm movements. "Painting" with water on a dark chalk surface is another way of mixing two familiar mediums. There's a touch of magic when the water evaporates and no messy cleanups when the fun fades.

## *PRETEND-PLAY*

Much of the pretend-play of toddlers is still geared more to action than to complex story development. But the beginning of a new and significant form of play does begin to unfold at this age.

"Go sleep," two-and-a-half-year-old Jennifer tells her doll as she plops her into the carriage and wheels her across the room. "Bottle?" she asks as if her baby could answer. This ability to play little dramas of pretend begins to blossom during the third year. Although young toddlers may imitate grown-ups by talking on a toy phone or sweeping the floor, the older toddler begins to play more elaborate games.

Making believe is more than just cute, or fun. It's a child's way of trying on roles they couldn't possibly fill in reality. Clunking around in Mommy's high heels or carrying Daddy's briefcase, the toddler steps outside himself and into a role of power and control—an unusual state of affairs for the relatively powerless child.

Through dramatic play, children learn a great deal about themselves and others. It offers a way to ventilate and overcome fears, both real and imagined. Jenny can't pinch her new baby sister but she can pinch her doll without getting into trouble. Bobby can't drive a real car but he can beep the horn and steer the wheels of his toy taxi or his train.

This ability to use toys "as if" they were real objects is unique to human beings. The young of many animal species play for a time. Kittens and puppies run and chase each other in rough-and-tumble games just as frisky colts and fawns gambol and play on wobbly legs. But playing games of pretend is an intellectual leap that demands a higher level of thinking.

For young pretenders, toys for dramatic play need to be realistic. The wheels of imagination are still hemmed in by the two-year-old's rather limited experiences. In time, children can transform a leaf into a boat that sails on a puddle

*Let's Cook*
*Galt*

sea, but for now a boat that looks like a boat helps to launch the imagination.

Young children begin their games of make-believe by playing out familiar roles and events. Since home and the family are the world they know best, this is the logical place for pretending to begin.

Housekeeping equipment and dolls are enjoyed by both boys and girls at this age. A small table and chairs are basic equipment for tea parties, art work, puzzles and even for snacks and mealtime. Look for a scrubbable table surface and chairs that don't tip. Avoid fancy three-legged stools or chairs that are too tall. Your toddler's feet should touch the floor.

If you haven't done so yet, invest in a sturdy set of toy dishes and add a few solid pots and pans. Like the dishes described earlier, Bambola's pots and pans come in bold colors—without cutesy licensed characters. Or consider Galt's metal pots and pans that look like the real thing. These are more expensive than the usual, but remember, these are playthings that will be used for several years. Unfortunately, most of the dishes and pots in toyland these days are undersized, unattractive, and made of flimsy plastic that snaps easily.

Since imitation is what they like best, twos enjoy child-size versions of household equipment Mom and Dad use. The Fisher-Price Bubble Mower and their Magic Vac are variations on that old favorite, the push-along toy. While a real vacuum cleaner or mower may be so noisy it frightens some twos, taking hold of a child-size version may help diminish such fears. In playing out roles children are often able to gain a sense of control over objects and situations that produce anxiety.

In many homes, permanent press fabrics have made ironing boards and irons anachronisms, but if ironing is something a child still sees, a toy ironing board and iron (with no plug) will be of interest.

*Magic Vac*
*Fisher-Price*

A doll for toddlers should be lightweight and simple. Dolls with printed or stitched on features are still safest. Be sure the doll's clothes and trimmings can't be swallowed. Forget about fashion dolls like Barbie for this age group. Soft dolls are more huggable and luggable, and at least one doll should be tubbable. Ideally tub dolls would have unjointed limbs so they don't fill with water and drip where they go next. Unfortunately they're not being made that way. Corolle does make a small Bath Baby with an all-vinyl body and bald head. It comes dressed in a bathrobe, and to empty the bathwater you pop off a limb. Effanbee's Butterball Baby or Softina from Goldberger Dolls work on the same principle. Dolls with hair are not recommended for this age group, nor are toy feeding bottles.

Your child will happily use a toy crib or cradle with small blankets and cushions for putting dolls and stuffed animals to sleep. You can buy one or make one from a box or basket. Also nice for strolls around the house or backyard, a small doll carriage is another kind of push-along toy with excellent pretend potential.

Most of the toys suggested here can be used over the next three or four years. No child needs them all, nor should they arrive all at once. Parents who are handy with tools can easily build some of them, such as a toy kitchen with a plastic sink tray, for a fraction of the cost. A less durable alternative is to paint cardboard boxes that are scaled to child-size.

Many of the best props for domestic play are household containers, empty packages, and hardware that children can safely handle. For instance, an old-fashioned eggbeater will whip up a mountain of bubbles in a play sink. A colander or tea strainer or ketchup squeeze bottle makes interesting drips and squirts, and a sponge is fascinating for squeezing, floating, and scrubbing.

For children of this age, some of the plush animals that arrived as infant gifts will come into their own. Aside from the special favorite, that huggable luggable that's carried off to bed or on trips away from home, there are others that are used as props for pretend-play. They take rides in wagons, sit up in the doll carriage, and often have to be put to sleep in a special order as part of a nighttime ritual.

A toy dashboard with steering wheel, horn, windshield wipers, and stick shift makes a wonderful toy for get-away trips when you can't leave home.

Dress-up clothes are less important and elaborate now

*Toddle Tots Family House*
*Little Tikes*

than they will be in a year. But older toddlers do like to put on hats, grown-up shoes, pocketbooks, and briefcases. Old jewelry is dandy and so is a fake wristwatch that looks like the real thing. A full-length unbreakable mirror is a great delight for seeing their favorite person and trying out funny faces.

Older toddlers also enjoy another form of pretend that involves manipulating small objects and playing multiple roles. For years the Fisher-Price Little People Environments have been recommended for toddlers, but we now know that the peg people represent a choking hazard, since toddlers are still likely to mouth their toys. The same is true of many wooden miniature farm sets. Play it safe. Save them for later. Little Tikes' Toddle Tots Family House comes with movable furniture and people that are large enough to be safe, yet small enough to be satisfying for the toddler's little dramas. Eventually these mini-worlds and their pieces can be used along with the block settings that preschoolers create. For now, they provide the bold outlines and realistic pieces for early fantasy games that are more manipulative play than fantasy.

## *Intrusions into Early Imaginative Play*

All too often this is the stage when TV's "imaginative" cartoon characters slip into your life. This is especially true of toddlers with older playmates or siblings. One mother told me that her two-year-old's favorite doll is She-Ra, He-Man's sister. What does she do with her? According to Mom, she feeds her, sings to her, and puts her to bed. It's some comfort to know that twos will be twos and use a toy in their own age-typical fashion. However, one can imagine a more suitable doll for such games of pretend. Falling into the TV-toy "buy-me's" at two is a mistake that simply sets the stage for more mistakes.

Pretend-play at this stage should be based more on reality

than on far-flung adventure. Young children have quite enough to deal with in sorting out their real world and real feelings. In fact, one of the things they're sorting out is what's real and what's make-believe. Most of the cartoons on TV and the toys they spin off simply don't enhance the child's budding imaginative play. Stay closer to earth for now. Kids need a firm footing in the here and now before they're ready for trips into the past or the future.

## *MANIPULATIVES AND PUZZLES*

A good many of the everyday things children use in the course of a day challenge their eye-hand coordination. Dressing and undressing themselves and their dolls, opening and closing boxes, doors, and faucets all require different kinds of fine motor tuning. In many respects everyday tasks are more interesting than toys specifically designed for repetitive skill building.

*Rollercoaster*
*Anatex/Kinderworks*

The expensive but fascinating Rollercoaster by Anatex and the Kindercolor Express by Kinderworks are practically identical. Either toy has multiple learning possibilities. It's a three-dimensional maze of colorful curved wires with beads in various shapes, sizes, and colors. Rollercoaster is like a witty piece of sculpture that keeps changing and stimulating language, eye-hand coordination, and exploration. Because toddlers are unpredictable and the wires may be inviting, this toy needs supervision for now.

Ring-and-post toys can now be sequenced in size order, and so can nesting cups with as many as five to ten pieces. Since toddlers can now turn toys that screw and unscrew, a set of kegs that fit one into another is interesting. Some of these come with a plastic toy in the smallest barrel, which presents a choking hazard. Remove it.

Simple four- or five-piece puzzles that fit in a tray are enough of a challenge to begin with. Those with whole objects that lift out are best for beginners. Also helpful now are puzzles with peg handles to lift. By year's end some children can handle puzzles with as many as twelve pieces. Although puzzles are expensive they are not for one-time only use. Remember that the two-year-old's love of repetition makes putting puzzles together again and again quite acceptable. More crucial is keeping stray puzzle pieces from getting lost. A storage shelf or puzzle rack should keep things where they belong.

Shape sorters with five or more shapes are now a welcome challenge for eye and hand. The older two-year-old will no longer try to jam any old piece into any old hole. He looks, thinks, and then acts. This is a new level of play, a more

*Busy Shapes and Key House*
*Playskool*

sophisticated approach to problem solving than the old trial-and-error method. You can talk about the shapes and colors, but don't turn the toy into a lesson. As we've discussed, knowing the name of an octagon at two will not get your child into Harvard.

Though children at this age are quite fascinated by keys, many of the key-fitting toys continue to be more frustrating

*Snoopy*
*Determined*

than not. Playskool's Busy Shapes and Key House have color-coded doors and big keys that simplify matters.

Though it will be some time before they can do the whole job, toddlers enjoy toys with snaps, laces, buckles, and zippers, just as they enjoy the challenge of a pair of shoes with buckles or laces. Dolls made for practicing those skills, like Dapper Dan from Fisher-Price or Snoopy from Determined, won't take the place of working on their own clothes, but are useful nonetheless. After all, it's a lot easier to refine those skills on a doll than it is on buttons, zippers, and snaps that are right under your nose.

Floor toys that invite manipulation with a rewarding payoff may require a little demonstrating. Playskool's Sesame Street Busy Poppin Pals features five puzzles. When the right knobs, switches, dials, and levers are worked, a little door opens and Big Bird or one of his friends pop up. In the beginning, your toddler may only be able to push the doors closed or activate one puzzle, but later on she'll know how to operate each mechanism and know just who will pop up. This is a toy that develops dexterity, problem solving, and visual memory.

*Sesame Street Busy Poppin Pals*
*Playskool*

*Bunny Builders*
*Lardy/International Playthings*

## *Blocks and Construction Toys*

Young twos will still enjoy the big cardboard blocks and the nesting blocks for stacking and hauling that I recommended earlier. Wooden kindergarten blocks are rarely of much interest before two and a half. Even then, they will be used more for manipulating than building serious constructions. Again, the two-and-a-half-year-old may call his tower the Empire State Building, but the name is given after the fact. He doesn't start out to build a particular building. That kind of imaginative block building will take off during the preschool years. For now, a small set of blocks will most likely be hauled in a wagon or loaded and unloaded from the shelf or box. Toddlers can make wobbly towers or knock down the ones you make, but blocks are just too abstract for most twos to use for dramatic play.

Plastic construction sets that snap, screw, link, or simply stick to each other all invite young builders to use their hands and eyes to work together. Lego's Duplo Blocks, Playskool's Bristle Blocks, or Bunny Builders from Lardy offer varying textures, sizes, and shapes for fitting together and pulling apart. Ritvik's oversized Mega Blocks, which look like giant Legos are also welcome now. Sometime during this year a stack of bricks or blocks will not merely be joined together, it will "become" a named object. "A bus!" Jonathan may announce. "See! A bus!" Even if the name is given after the fact, the ability to make a stack of blocks stand for something else is another intellectual achievement that will blossom during the preschool years.

## *BOOKS*

Though books are technically not toys, their entertainment value is related in several ways. From the child's point of view the mechanics of manipulating a book is definitely a toylike experience. Indeed, many books intended for older toddlers enlarge upon this mechanical quality. Eric Carles's Spot books with lift-and-look surprises incorporate a variation of peekaboo and hide-and-seek. There are books with textured and scented pages. Others have wheels, handles, and interesting shapes. For independent "reading," cardboard pages are still desirable. Twos are fond of poring over books with clear and familiar pictures. They like the rhythm and bounce of verse as well as simple stories that reflect their own experiences.

Sharing books together after dinner or before bedtime is an excellent way to provide a change of pace. Twos enjoy many of the I Spy games suggested for younger toddlers. And they enjoy chiming in on stories with a repeated refrain. In fact, one of the things twos love most about storybooks is their predictability. They love hearing the same story again and again. Just try changing a line and you'll get the message.

Many books are tied in to licensed TV characters, whose presence in toyland is inescapable. Although the quality of many such books is negligible, the familiar characters on their covers guarantee instant recognition and easy sales. As a result, most children probably have their first experiences with books purchased in the supermarket or toystore.

There's nothing wrong with an occasional impulsive purchase in such a setting; nor are all mass market books badly written or illustrated. Aside from being inexpensive, they are a reasonable way to reinforce the child's desire to connect with books and familiar TV characters. But they should not be the only source of reading material. For twos and up there are definitely better choices to be found in bookstores and public libraries. For guidance along these lines check with your local library or see Bank Street's *Choosing Books for Kids.*

Wherever you shop, remember that the books you buy are likely to be memorized word for word and read again and again. So choose carefully, paying close attention to the language and the illustrations.

## *STORAGE AND SAFETY*

So many of the toys for this age group come in multiple pieces it's small wonder that playthings often clutter up every corner of the house. Unfortunately, toy chests are basically useless pieces of equipment. They're usually too deep for toddlers, and hazardous; the toys generally end up dumped in a jumbled heap with pieces lost or buried. Rather than a toy chest, why not organize their toys with an eye toward independence and learning power?

Low open shelves for large toys are both decorative and accessible. For toys with small multiple pieces, the box they

*Warning: Old-fashioned toy chests with free-falling lids have caused terrible accidents and even fatalities! If you have one, remove the lid until you can install safety lid supports. You can order them at very little expense from the following companies:*

*Carlson Capitol Manufacturing Company*
*P.O. Box 6165*
*Rockford, IL 61125*

*Counter Balance Supports Company*
*4788 Colt Road*
*Rockford, IL 61109*

*You must include the dimensions and the weight of the lid in your order.*

come in will rarely go the distance. Better to set up plastic carriers or baskets that will go on a shelf or on the floor. A basket or see-through plastic box can be provided for small cars and a different box for miniature animals. Putting toys away or taking them out will eventually become a little exercise in sorting and categorizing. It plugs right into the toddler's own love for order and things having a place where they belong. Providing the framework for that kind of thinking won't guarantee a neat house and no arguments about picking up, but it can make cleaning up less of a hassle and more of a game.

Because broken toys can be a real hazard, be sure to check the toy supply from time to time. Dispose of toys with lost pieces or broken parts. Pass along the toys your toddler has outgrown to a younger child or pack them away for a future toddler. Avoid a buildup of clutter that gets in the way and ends up clogging children's play. Everything doesn't need to be out and on display. Toys that are out of sight for a while are like newly discovered friends when they're reintroduced.

## THE PARENT'S ROLE

Buying, making, or providing playthings is just the beginning of encouraging your toddler's play and learning. Though two-year-olds are increasingly able and eager to go from one activity to the next on their own, the value of some shared play should not be overlooked. A little one-on-one time offers the opportunity to talk and laugh together, and to try new ways of using play materials and ideas.

You don't always need to sit down and stop what you're doing. While you're loading the dishwasher or folding the laundry you can "call" your toddler on the phone and ask him to deliver some food for lunch. If she's riding around the kitchen on her pony, offer her some make-believe carrots to feed it or send her on an errand, real or imagined. As children develop the ability to pretend, they will need fewer cues, but for beginners a little modeling goes a long way.

They may need help with new toys, too. A new puzzle is much easier to take apart than put together. When your toddler gets stuck, try not to take over. Offering a few hints supports a child's wish to "do it myself" while minimizing her frustration.

A set of blocks or a miniature farm or garage will be emptied and investigated piece by piece. Allow time for this initial exploration and freedom to use the pieces in the child's own way. Take your cue from your child as much as possible. You may need to demonstrate how pieces fit together or suggest experimenting to see if the cow can fit in the barn or a car will go up the ramp. On another occasion you might want to sing "Old MacDonald Had a Farm" and have your toddler find each animal you're singing about. The object is not to turn such shared moments into a drill with right and wrong answers. By all means talk, but let the learning flow from what the *child* wants to know or do, and keep it playful.

Although few toys or activities will hold the toddler's interest for long stretches, some have more staying power than others. Many art materials are especially fascinating even after the first flush of novelty wears off. Since most art supplies call for some supervision, you can't just put them out and say, "Don't make a mess." You can, however, set things up and get the action started. Once children know how to handle paints or play dough or chalk themselves, you can simply be nearby doing what you need to do. From the sidelines you can comment about the lovely red dots or the bright blue lines. Or you can grab a brush and do a bit of painting yourself. The thing is to be there to get things started and to keep them going if the child get stuck.

A great many toys are noisy and messy, or require supervision. If you can't stand banging and thumping, don't bring home a hammerboard or drum, or you'll end up spoiling the purpose of giving a toy in the first place. A climbing device with a slide will be used for physical as well as pretend-play. But if you haven't the time or patience to supervise young mountain climbers, they'd be better off without it. *Know your own limits*—but try to expand them if you can, as children love noise and mess a great deal more than adults do.

## SUPPORTING NEW SOCIAL SKILLS

For twos, a trip to a playground or park is not merely physically satisfying, it's socially stimulating. Sitting in the sandbox, climbing on the slide, riding back and forth on a swing are all grand. But motion is just part of the picture. Toddlers are also attracted to other children, whom they like to watch and imitate, at first from a distance. Before the year is out though, twos who have opportunities to play near other children will be playing *with* them, not just next to them.

These early steps into an expanded world of playmates herald a new kind of social development. Children of this age have much to learn about the give-and-take of cooperative play and sharing toys, and their readiness for playmates develops along with their ability to communicate.

As language ability and experience grow, so will the child's ability to negotiate with others. The desire to play will eventually outweigh the need to keep every toy just for himself. For beginners, however, parents and other caring adults may be needed both to negotiate and to protect the child from his own lack of social savvy.

Much of the push, shove, and grab of toddlers' play can be limited when adults are on the sidelines to support the equal rights of the grabber and the grabbed-from. With very young and inexperienced players, though, the path of least resistance is to have more than enough toys to go around. Sometimes the offer of an alternate scoop or shovel will satisfy the child who is more concerned with what's "Mine!" than about the object in question. "When Johnny's had a chance to fill the pail, then he'll give you the shovel" may oil the wheels of taking turns, but not always.

In the sandbox, one pail and one shovel won't satisfy two kids. A certain amount of duplication is necessary; it makes sharing a sieve, sandwheel, or dumptruck easier. Nor can you expect toddlers to learn quickly about the nuances of give-and-take, or of leading and following. Frequent but brief social experiences—an hour at the park, say—are better at this stage than long and pressured playtimes. By late in the second year, the time may easily stretch to an amicable hour and a half or two. Remember, in shaping positive attitudes, the quality of time together may be of more value than the quantity.

Until recently, few two-year-olds played with children outside the family circle. But as more women have entered the work force, more young children are being placed in daycare settings. Preliminary studies show that young children who spend their early years in group settings do not merely play side by side with minimal interaction, but are capable of showing considerable interest in one another. Though some of that interest is expressed in aggressive behavior, children are also likely to express concern and support for an agemate who needs comfort or attention.

Whether your child is in daycare or at home, he should also have some space and time for solo play within the group setting. Kids need time to watch as well as join in, to play near each other as well as together. Adult caregivers need to be available as resources, but must not constantly direct the child's play. Toys that can't be used independently are for the most part a poor choice for twos. If most of a child's toys are too difficult, they can rob the child of his self-confidence.

For parents who are at home and find themselves and their toddlers cut off from other parents with children, a good bet may be to organize an informal playgroup that meets for an hour a few times a week. The benefits of socializing with others may be as rewarding for the adults as it is for the toddlers. Initially all parents will need to stay close at hand while the toddlers get to know each other. But eventually one parent may be able to handle a few children while the other parents use the free time to do chores or simply relax. Taking turns supervising gives everyone involved a little time off while the toddlers learn that there are other caring people in the world they can depend on and enjoy. And remember that the size of the group should be limited as well as the time.

Whether there are two, four, or even more toddlers in a group, forget about everyone playing the same game. Few twos are ready for big group activities. Certain common materials and playthings should be available, but that doesn't mean everyone will use them at the same time. For children of this age, one-on-one socializing usually works best. In other words, with twos, a duo is more apt to play together happily than a trio or a quartet.

Materials like play dough and crayons are more easily shared than a puzzle or a single truck or doll. A sandbox with plenty of containers and scoops can be happily used by several children whether they're interacting or simply playing side by side.

## *Making Choices*

This is the age when licensed characters often begin to attract attention and start dominating the toy choices of children and parents.

Looking for toy dishes for my granddaughter's second birthday, I found Mickey Mouse, Big Bird, Cabbage Patch Kids, and My Little Pony plastered on every plastic plate, cup, and teapot. In the end I selected a different license, a set of Corning Ware toy pots and dishes. At least they had some connection to the real thing—rather than a connection to TV.

Searching for "clean" dishes reminded me of a similar experience at the Toy Fair several years ago. In looking for a sled for the toy article I was writing, I couldn't find one without a licensed character. In the end I surrendered to the Smurfs. There was some satisfaction in knowing children were going to end up sitting on the jolly little guys.

The point is, sometimes there are no better choices, but often there are. Usually the license is not intrinsic to the toy. Given a choice, is there any reason to buy an easel, a trike, or a toy kitchen that's jazzed up with My Little Pony, Rainbow Brite, or Big Bird? Do they add to or distract from the play value of the toy? In buying such toys, are you also buying into a play style that leads to more of the same?

If you don't want to get caught in that trap, this is the time to take control. Where there are choices, opt for fewer logos and TV tie-ins. Without a doubt your toddlers will see and want some of the familiar icons. Buying *some* is not the problem—the problem is buying almost *everything* with one license or another.

Before you buy, ask yourself:

- Is it the license or the toy my child is attracted to?
- Does the character fit the toy or is it just stamped on for instant recognition?
- Is the toy beneath the logo really worth buying?
- Would I buy the toy if it had no license?

## *How Smart Are Early Smarts?*

Pressure for early learning and achievement often becomes intense at this stage. During the past two decades the curriculum of early childhood has been taking aim at younger and younger children. The pressure comes from many sources, including toyland with a deluge of supposedly educational toys.

While visiting a father-child playgroup recently, I watched a two-and-a-half-year-old overachiever and his father in action. The boy did not join the dramatic play activities or use the paints, blocks, or water that involved all of the other children. He talked only to his father and played only with a complex shape sorter. Indeed, the talk between them sounded like a test or a performance. The father held up geometric shapes and asked, "What's this?" The child named not just one or two familiar shapes, but everything from triangles to octagons. When the child started to stack the pieces instead of fitting them into slots, he was told, "That's not what they're for." The lesson ended when the child swept all the pieces off the table in an exasperated gesture of defiance.

The fact that a two-and-a-half-year-old can recognize and name an octagon may signify some precociousness; but how important are early smarts, and to whom are they important? All too often, early achievements like these show that a great deal of parental time and pressure have been expended in teaching, often at the expense of active, age-appropriate play.

Parents have always been eager for their children to learn. But the content of those early lessons has become more formal and narrowly defined, with heavy emphasis on school-related skills. Academic skills no longer wait for entry to elementary school. In some places knowing your shapes, colors, ABCs, and one-two-threes has become a prerequisite for entrance to the "right" school. Indeed, in large cities many two-and-a-half-year-olds have to be ready for the academic treadmill of applications, interviews, and tests in order to win a coveted place in the "right" nursery school. Acceptance means they're on the track that leads to the right elementary school, high school, college, graduate school, career, success, and money. The fact that some two-and-a-half-year-olds, both those who "make it" and those who don't, get bent out of shape on the fast track and carry the scars as extra baggage seems to be a risk many parents are willing to take.

It's not that children of two are too young to learn or are uninterested in learning. One of their chief delights is in learning to name the world. But, they are also learning bigger lessons about learning itself and developing long-lasting attitudes about taking risks. Too many toys that emphasize intellectual lessons hammer away at counting, colors, words, and shapes. Such an insistence on facts and labels can short-circuit the joys of learning and play.

## *SUMMING UP*

In buying toys for twos, keep these questions in mind:

- Does my child have toys that encourage active physical play?
- Does my child have playthings that can be used for social and solo play?
- What toys have I provided for pretend-play?
- Will my child be able to use this independently and safely?
- Is my child ready or am I rushing to teach too much too soon?
- Is the toy something that "plays itself" while the child watches, or will the child do the playing?

# V. *Toys for Preschoolers: Three to Five Years*

## *THE NATURE OF PRESCHOOL PLAY*

What clothes and records are to teenagers, toys and play are to preschoolers. Even if they've never seen a TV commercial, there is probably no age group more attuned to toys. Try to imagine a holiday or birthday for a three- or four-year-old without them. Indeed, whether preschoolers are eating, dressing, bathing, or simply going to bed, there's apt to be some play involved.

For preschoolers, play is not merely fun. It's through their play that young children continue to learn about things, people, and, most especially, about themselves. At this age there are several developmental changes in the nature of their play.

## *The High Drama of Make-Believe*

First and most notably the preschooler's play comes alive with drama and action. When he's climbing the monkey bars or pedaling a trike, there's more going on than pure physical action. "Giddyap!" the young cowpoke calls as his rocking horse trots off wherever he commands. Unlike the toddler whose pretend-play was largely imitative, the preschooler's imagination sends him on more freewheeling adventures.

Preschoolers don't just imitate the adults in their lives. They also invent games of make-believe that are complex and dramatic, embellished not just with action, but with a story line as well. As a toddler, Jenny ran her vacuum cleaner just like Mommy; now she will scold her dolly for leaving crumbs on the carpet and deliver strong words about being a bad girl as she goes about her cleaning.

Dressed up in Mommy's beads and high heels or Daddy's hat and tie, preschoolers can step outside themselves and try out the roles of power that are denied them in reality. Indeed, through dramatic play the preschooler can transform herself into a growling tiger, a daring superhero or a helpless kitten. These flights of fancy offer children a way of dealing with feelings and fears. They are an outlet for tension, for fulfilling impossible wishes, and for rehearsing a broad range of emotions.

*Trike*
*Hedstrom*

## *Reality and Fantasy*

For the preschooler, the line between reality and fantasy is still tentative. The three-year-old is not at all certain that last night's dream didn't (or won't) happen. He's also not sure that wishing can't make things come true, or that others can't see the angry thoughts he sometimes has in his head. Although he's eager to please the adults in his life, it's not always easy to do what they say. In dramatic play, preschoolers can act out some of the conflicting emotions they are feeling. Sara can't pinch the new baby who takes up so much of Mommy's time, for example, but she can spank the doll who's naughty and won't go to sleep. Young John, meanwhile, can't fight off the monsters who lurk in the shadows, but dressed in his Super Hero cape he becomes the powerful master who triumphs over evil.

While in reality the preschooler is subject to rules dictated by parents and teachers, in play she becomes the rule maker. Like a conductor who lifts her baton, the preschooler says, "Let's play monsters," or "Pretend I'm the Mommy and you're the baby." When the going gets rough the child-player can stop the game or change the direction of the story line. Play has more options than reality, and it's the child who is in control.

Playing games of pretend is more than something cute that little children enjoy. It is the child's own way of testing and developing the boundaries of fantasy and reality—another major developmental task of the preschool years.

## *Where Do Fairy Tales and Fantasies Fit?*

Unfortunately many children's playlives are shaped predominantly by TV cartoons rather than their own fantasies. And TV's premade fantasies and toys play on narrow themes that don't bear much resemblance to the preschooler's world of here and now.

Producers liken current TV shows to classic fairy tales with their struggles between good and evil. Overlooked entirely is the fact that fairy tales are hardly considered appropriate fare for children as young as two, three, four, or even five. Although Bruno Bettleheim, in his book *Uses of Enchantment,* popularized the notion that classic fairy tales have value in providing emotional grist for children's inner struggles for independence, he wasn't talking about preschoolers. Nor did he advocate anything in the way of pictorial representation. Indeed, publishers who rushed to issue new and lavishly illustrated volumes of *Sleeping Beauty, Hansel and Gretel, Beauty and the Beast,* and so many more fairy tales, entirely overlooked Bettleheim's view that the stories are best told *without* illustrations.

Why no pictures? In listening to or reading a story without illustrations, the child makes his own images from his own feelings and experiences. Without premade images, the child is free to cast himself or anyone else, real or imagined, in any role he pleases. He can bc a cruel giant or a lost prince fighting off a magic spell. For the middle-years child (ages six to twelve) fairy tales are soul-satisfying tales of adventure. Danger is faced with courage, and the young, inexperienced hero is able to overcome the forces of evil and sundry other obstacles along the way. To schoolage children, such far-flung fantasy may offer a meaningful metaphor for their own journeys toward independence. Unlike their preschool brothers and sisters, the middle-years child has a grasp on the boundaries of what's real and not real. They can suspend belief and relish the spine-chilling dangers, knowing that such things only happen in "make-believe."

To liken today's TV shows to fairy tales is like comparing watching TV news to reading the *New York Times.* True fairy tales come alive with complex plots, rich language, and implicit lessons about a wide range of human experience and emotions—not simply good versus evil. TV's action adventures, on the other hand, leave little to imagine visually or thematically. Compared to classic fairy tales, the plots, characters, and language of cartoons are of the most simplistic level. The fact that preschoolers can follow such stories is not surprising, but does that make the stories or the toys they spawn appropriate?

For preschoolers, the predominance of such entertainment remains questionable. It is the real world and his place in it that the preschooler is trying to understand. Even as he becomes a growling tiger or a stomping monster, the roles he takes on have a transparent resemblance to his real-life wishes and fears. At this stage it is reasonable to believe that children have enough to handle from their own active imagination without heavy doses of pseudo–fairy tales that play on narrow themes.

Dealing with TV becomes a serious issue at this stage. There's a temptation to use the tube as a built-in babysitter —a sitter that kids seldom object to and are all too eager to rely on.

The fact that preschoolers are fascinated by the excitement on the screen does not make the content appropriate. In fact, young children often pay closest attention to things that confuse and frighten them. But sometimes we simply can't give children everything they want. After all, kids like ice cream and cookies, but we don't serve them all the sweets they'll eat. Instead we try to balance their diets. In the same way, parents need to play an active role in controlling TV and toy choices.

TV's action figures may dominate the toy industry's sales figures, *but they should not dominate the playlives of preschoolers!*

## *Play and New Social Development*

When she was a toddler, your child's interest in other children was more exploratory than social. Now her world expands to include others beyond home and family. Preschoolers don't just talk *at* other children, they talk *with* them. Indeed, before long, they will be talking more to children than to adults.

Although preschoolers are essentially egocentric and find it difficult to see things from another person's point of view, through play they begin to de-center—to learn that there's a "yours" as well as a "mine," and a "we" as well as a "me." So play becomes a vehicle for social growth. Though this morning's best friends may be mortal enemies by afternoon, the idea of friendship is beginning to germinate.

At this stage, parents and teachers are still needed to mediate or model ways of negotiating small disputes or heated fights. For the still-inexperienced player, the idea of fair play can be tricky. Controlling the impulse to grab, shove, or punch someone isn't easy, especially if someone has just knocked over your tower of blocks or called you a nasty name. For preschoolers, the old adage, "two's company, three's a crowd," generally holds true. Although preschoolers begin to find some group experiences enjoyable, they can usually sustain play better in pairs than in trios or larger groups.

In the company of other children the preschooler gradually learns about sharing and taking turns. In fact, the desire to play with agemates comes to outweigh the need to have everything his own way and helps him accept the necessity of giving as well as taking.

## *Playing with Ideas and Words*

Compared to the toddler, the preschooler can turn the most ordinary activity into a game. Send him to wash his hands and he gets caught up with producing a cloud of sudsy bubbles that are too fascinating to rinse away. Ask him to put a box in the garage and he becomes He-Man, able to lift a mighty boulder and ward off the forces of evil. If you leave a four-year-old to pick up her Legos, she may start building anew or become totally immersed in sorting the pieces by size or color. By five, she will be able to sort and match using more than one attribute. Preschoolers are not just more imaginative, they are more inventive and playful. They play with objects, ideas, and words—anything that comes their way.

Although preschoolers understand most adult conversations, and their own use of language is growing, they vary in their ability to make themselves understood. Play not only facilitates the development of language but also provides an arena for giving expression to the feelings, fears, and problems which children are unable to express in words.

Children also use language to teach themselves right from wrong in an attempt to internalize their parents' rules. This is the age when imaginary playmates appear, or a teddy bear takes on the role of scapegoat. That way it's not the preschooler who broke the tea cup, but his naughty invisible playmate. In creating this dichotomy of good and bad child, the preschooler becomes his own teacher and critic. He also takes a few steps forward toward another significant developmental goal—the forming of a conscience. Of course, it will be several years before the child has really internalized the controls and can do more than parrot the values of the important people in his life. But in controlling his imaginary playmate, he's both teaching and learning the broad outlines of right and wrong.

Unlike toddlers, who tend to go from one activity to the next, preschoolers have more staying power. They're able to plan ahead, to combine ideas, and think of new ways of using familiar objects. If Jon is building a barn with his blocks, he will put a horse, not an airplane, in the barnyard. His pretend-play has a logic learned from reality. If he needs to go downstairs to find the horse, he is not likely to be diverted by whatever attracts his attention. There is a new kind of seriousness and purposefulness to his play.

## *More Than ABC*

Kids of four and a half don't need learning games to teach them to read at the third-grade level. Don't feel guilty about not buying your three-year-old an electronic Dial-a-Teacher that's programmed for the way three-year-olds think. Threes need to be allowed to think like three-year-olds. They don't need their plastic construction toys plastered with letters or addition and subtraction symbols. Until they're ready for crossword puzzles, learning to spell vertically is not a good starting point. Yet that's how it's done on Selchow and Righter's Scrabble People, Ritvik's new Mega Tech Blocks, and Tyco's Preschool Building Set. Why? So they'll be perceived as educational. Forgotten completely is the fact that the child's own constructions are a significant symbol system with more basic and meaningful learning value at this stage.

Although many of the so called "educational preschool toys" put heavy emphasis on school-related skills, the young child's learning style can't (or shouldn't) be limited to such compartmentalized lessons. Indeed, too much emphasis on solely cerebral exercise may rob preschoolers of the multiple kinds of learning experiences that flow from active and interactive play.

In our eagerness to provide children with stimulating learning materials, we should be relieved to know that half an hour in the sandbox is no less educational than moving a set of magnetic letters and numerals around on the refrigerator door. Nor is fitting a complex puzzle together more important than riding a trike, playing house, or building a tower of blocks.

For the preschooler, play is a vehicle for much broader and important lessons than learning how to count or say the ABCs.

Consider the multiple kinds of learning the preschooler experiences digging in the sandbox. First of all, sand lends itself to such a variety of sensory possibilities. And you can pour it, pound it, dig it, mound it, run it through a sieve, or scoop it with a spoon. If it's dry it runs through your fingers. Wet it and it turns to mud that molds into shapes. All of these rich sensory experiences are the underpinnings of meaningful concepts and language development. Words that are basic to an understanding of spatial relationships, science, and mathematics are learned easily in the context of such play. Concepts like more and less, wet and dry, light and heavy, over and under, full and empty, and smooth and rough are just some of the practical things learned in the sandbox.

### *Unstructured Materials*

For the preschooler, unstructured materials like sand put her at the center of her play. She makes the material come alive by acting on it. If she does nothing, the sand does nothing. It's what the child does that shapes the play. The ideas spring from the child rather than from the material. In fact, as the child plays, a mud pie that is a birthday cake one moment may become a mountain for monsters. An unstructured material like sand has a flow to it, just like the child's own imagination. With such a material, there is no right or wrong way to build a roadway for a truck or a lake for a boat.

### *Symbolic Thinking*

In playing with sand, paint, water, clay, or building blocks, the preschooler makes her own structures, her own rules, her own vision of reality. This new ability to transform raw materials into whatever the child wishes is no small intellectual feat.

In transforming a pail full of wet sand into a birthday cake with twigs for candles, the preschooler is playing with symbols of her own making. In similar fashion she can roll a ball of clay into a long wiggly coil that becomes a snake, or run a block across the floor and turn it into a jet taking off for adventures. Just as she can transform herself into She-Ra, she can transform objects, making one thing stand for something else. This new ability to create her own symbols is closely related to the kinds of skills she will need when she uses more abstract symbols like letters and numbers.

## *The Push for Early Achievement*

Some parents like to buy "advanced" toys to prove what "smart" progeny they have. "Janie loves this puzzle even though it says it's for six-year-olds," a proud parent boasts of her three-year-old. Never mind that Mama is the one who's doing the puzzle, not Janie. Or consider Jason, a bright four-year-old whose parents have taught him to read. While visiting our home, Jason recognized a game that featured a favorite TV character. He read the name of the game and insisted on playing. "Let's read the rules," his dad said, pointing out that the game's label said it was for ages eight and up. Willingly Jason read each and every rule; but when it came to playing, Jason had his own ideas. Like any four-year-old, Jason could not deal with the complex rules and strategy required in playing a game well beyond his understanding. Rather than adapting to the boy's needs, the game was abandoned. "We have to play by the rules," his father said sternly, "or else we can't play."

For children who have to live up to unrealistic expectations, the rich values of play may be lost at an early age. Unfortunately, the push for early achievement is rampant not only in the realm of schoolwork, where parents push their children to read a "harder" book or to do next year's math workbook, but also in toyland. Lost entirely is an understanding that growth can be measured more readily by the way children play than by the age labels on a box. Watch a toddler with blocks and you'll see plenty of dumping and filling action. Then watch a four-year-old with the same blocks and you'll see towers and roadways and bridges. Switch to a six-year-old and you'll see a majestic castle being erected with turrets and towers, and a story being enacted with action and drama.

It isn't the toys so much as it is the child's developmental level that shapes the play. We often forget the pleasure children derive from having total mastery over a relatively simple toy. One has only to watch a preschooler playing at a shape sorting puzzle with a younger child. They relish not only showing the baby, but showing themselves how well they can do it. Like young readers who love going back to read and reread "easy" books, children don't always need a new and harder challenge. They also thrive on repetition and ironing out all the rough spots. It's only after they completely master the basics that they can go and use an old toy in new ways. Rather than always rushing to a new and tougher challenge, parents would do well to let children invent more challenges of their own.

But before preschoolers are ready to understand that the little black squiggles on a printed page are symbols for the words we speak, they learn a great deal by playing with symbols of their own making. This ability to imagine, invent, and transform objects is a new and important way of thinking. Children who have the rich and varied play experiences that support symbolic thinking, are better prepared to make the leap into reading and writing. They bring a wealth of knowledge—a process—that outweighs the little compartmentalized lessons which so many teaching toys hope to convey.

## *The Social Dimension*

Returning to the sandbox and its multiple learning experiences, let's not overlook its social dimensions. In the park or at the nursery school there are going to be other children. They may watch one another, imitate each other, or actually get into a cooperative venture like building a road.

It's in the sandbox that preschoolers find a common ground for expanding their social horizons along with their dexterity, language, and intellectual and imaginative development. Few play materials touted as educational cost so little and lend themselves to such rich and varied learning experiences.

But basically, I am using the sandbox as a metaphor to describe how free and creative play serves the child's best interests. In an age when early achievement is being sold to parents, it's important that parents understand just what kinds of play and playthings match the child's developmental stage and learning style.

## *Personal Choices*

It's during the preschool years that children begin to want a say in choosing toys. Those who watch a lot of television tend to ask for a lot of inappropriate toys. During the holiday season it's hard to keep track of their wish lists. Even children who watch little or no TV pick up on the current pop toys. Rather than turning a deaf ear, or, on the other hand, going out on a buying spree, consider their requests with an open, but critical, mind. Take time to look at the commercials, or better yet, go to a toystore and examine the toys. Consider how and where a particular toy will be used. Can it be used in multiple ways? Is it a toy that plays while children watch or does it invite children to do the playing? Is the toy well made or will it fall apart quickly? Can it be used solo or is it a toy that requires another child or adult participation? Are you prepared to devote some time to it? Is the toy messy, noisy, or otherwise intrusive? Can you deal with the messiness or is the toy going to be off-limits most of the time? Is the toy just another variation of several others your child already owns, or will it truly expand your child's play repertoire?

Often, the best loved toys are those that extend a current theme of play. For instance, a child who's busy with dolls will not necessarily find a new doll as useful as a high chair or stroller or dolls clothes and accessories.

Your child's play interests and toy choices reflect his or her individual style. Some children are more active than others. Some are more mechanically inclined. These individual differences are real and shouldn't be overlooked. Whatever children's personal interests, they should be given toys that invite them to stretch their playing repertoire. Preschoolers should have toys that encourage them to use their muscles, minds, hands, eyes, and imagination. They should have toys for quiet play, for solo play, and for social get-togethers.

## *HOW PARENTS ARE HANDLING VIOLENCE*

Violence on TV and in toyland has recently triggered a storm of controversy, but the topic is hardly new. Parents have been debating the issue of toy guns, soldiers, and armed warriors for generations. Some dismiss the current crop of good and bad guys as harmless. Here are some of their remarks:

> "My boys have Transformers and Gobots. They had Star Wars and He-Man and GI Joe. I guess they have them all. I don't see anything wrong with them. I used to play cops and robbers and war. How is it different? They're just having fun. I don't think it hurts them."

> "I always wore my cowboy suit and holster to the Thanksgiving family feast at my grandfather's. I didn't grow up to be a killer! I don't even own a gun. So how did it hurt me?"

By contrast, other parents wonder:

> "How can the same people who marched for peace in the sixties buy their kids all these war toys? What are they thinking?"

> "I'm appalled by the bunker mentality in toyland, and I refuse to finance an arsenal of war toys. My children know I work for peace. I refuse to buy toys that promote war. It's as simple as that, and my children accept it."

Between both extremes—those who forbid and those who permit—a greater number of parents have mixed feelings. An informal survey of parents of children enrolled in the Bank Street School indicates that, at best, the majority "abide" many toys. Although many of the parents surveyed forbid toy guns, several of the same group allow water guns in the bath or at the beach. Others allow guns so long as they "aren't used to shoot at anyone." Still others draw the line by separating reality from fantasy. They outlaw realistic guns but permit mythological weapons like magic swords.

The same sort of uneasy truce exists with action figures. The majority of parents in the Bank Street Survey forbid Rambo and GI Joe toys but allow Transformers, Gobots, and He-Man toys. "We're opposed to war toys but in favor of myths. He-Man doesn't kill anyone," said one parent. Many parents share the view that Transformers and Gobots are acceptable fantasy figures. "They're interesting kinds of mechanical puzzles." But even with regard to robots, many parents disapprove of those with elaborate weapons.

In our eagerness to find clear-cut answers, it's difficult to accept the fact that some issues are too complex for a simple yea or nay. Before taking an "anything goes" or an "absolutely never" stance, let's look at what we really know about kids and violence.

## What We Know and Don't Know

Research from NCTV (National Coalition on Television Violence) shows that the average American child will see eight hundred ads promoting war toys in a year and watch two hundred and fifty war cartoons that promote these toys. That's the equivalent of twenty-two days of classroom instruction.

We know that children's behavior is more aggressive after viewing shows with violent themes. Although researchers can only speculate on the long-term effects, there is no question about the here and now: shows and toys that feature violence lead to games of violence.

Although no research has shown that violent play in childhood leads to a life of crime or criminal tendencies, children's immediate reactions to violent programming cannot be ignored.

One of the most detailed studies of preschoolers indicated that children who viewed violent cartoons played more violently than children who watched Mr. Rogers or Captain Kangaroo. In more than two dozen other studies, researchers found that cartoon violence increased in children such behavior as hitting, kicking, choking, throwing, pushing, holding other children down, hurting animals, and selfishness.

One study has concluded that there is a correlation between viewing violence and increased anxiety. Heavy TV viewing was found to lead to a belief that the world is a "mean and scary place."

## Does Banning Work?

Considering the research, it's hard not to conclude that the logical answer for parents is to ban both TV programs and toys. Yet that may be the least effective way of helping children handle the issues of violence, aggression, and consumerism. Indeed, banning has a way of making things even more desirable. When dealing with children, it's important to use absolute prohibitions only on a limited basis.

Anyone who has lived with children knows that they can turn even the most benign plaything—including their own fingers—into a "bang-bang-you're-dead" weapon. In other words, even if you don't give your children violent toys, they'll probably invent them anyway.

We also know that play can be an outlet for releasing aggressive tensions. When a new baby arrives, big brother or sister may one moment emulate Mother's gentleness with a baby doll and use the same doll as a punching bag the next. As safety valves, toys and play can provide satisfactions that would not be acceptable or possible in reality. In fact, this may be why He-Man and She-Ra seem to satisfy young children's fantasies.

Some parents of four- and five-year-olds claim that their preschoolers get a real sense of power in their fantasy play with He-Man. It may be that the all-good, all-bad simplicity of the cartoon superhero matches the young child's simple view of right and wrong. Developmentally, this is the way young children sort out the world. Things are either pretty or ugly, delicious or yucky, mine or yours, good or bad, right or wrong. As one parent put it, "At least He-Man and She-Ra are 'good guys' with principles."

Indeed, some researchers have found that despite the violence and weapons, children who watch cartoon fantasy shows tend to imitate aggression less than they do after watching more realistic shows: "Miami Vice" triggers more aggressive behavior than "Masters of the Universe." However, for some preschoolers neither show would be appropriate and banning is in order.

Without totally banning TV or surrendering to its seductive power, there are things parents can and should do:

- Establish firm limits on the amount of TV watched and the number of TV toys bought.
- Consider the content of shows selected. Sit down and watch the shows with your child.
- Talk about what's happening on screen.
- Observe your child's reactions during the show and the kinds of play and acting out that follow.
- Provide alternative forms of entertainment. Outings, books, games, art supplies, toys, and video cassettes are all good choices.
- Limit your own viewing habits. Remember, children learn more from the example you set than from what you say.

## *Kids and Commercials*

In the Bank Street Survey we asked children what they thought about TV commercials? Four- and five-year-olds were essentially enthusiastic and uncritical. The only qualified response came from a four-year-old girl who said, "I hate boys' toys but I like girls' toys."

Although older children expressed their disappointment and skepticism about toy ads, preschoolers still seemed to believe that what you see is what you get. Indeed, the "getting" part frequently causes great friction between parents and children. As one parent remarked, "He wanted everything that he saw advertised until we made a rule: TV goes off if he says 'I want, I want!' "

While researchers continue to debate whether children know the difference between genuine ads and the toy-driven shows, there seems to be no doubt that for preschoolers the line between the two is meaningless. Children under eight don't even understand that sponsors want to sell products. We do know that children who watch a lot of commercial television request more toys. One study even indicates that children who are heavy TV viewers say they'd rather play with a "not so nicc" child who has a desired toy than with a "nice" child who doesn't.

Since preschoolers are so vulnerable to TV's multiple messages, parents need to take a more active part in setting limits to both viewing and buying. They also need to mediate the messages. Parents should point out not only what the small print says—"batteries not included," "some assembly necessary"—but also the bigger deceptions. For the preschooler, it's hard to understand that the music, the action, and even the child on the screen don't come in the box. Some parents find the best course of action is going to look at the real toy. Others say that they watch to see how much their children enjoy a new toy when they encounter it at friends' homes. In other words, their kids don't need to be the first kids on the block to own a "hot" new toy.

The key is to take control of the situation. You don't have to boycott everything that's advertised; but you can make decisions about toys on an individual basis, from toy to toy, and from child to child. By becoming selective buyers parents not only resist the domination of the best-promoted toys, they also have a chance to discover the value of some of the least-promoted and most important playthings for preschoolers.

## *TOYS FOR THREES, FOURS, AND FIVES*

While no child needs all the toys listed in this chapter, here are some of the best you can choose from. Since play is so central to preschoolers, keep in mind that these are the years when a child may need more toys than ever before—or after.

## *Dress-Up Clothes*

For their dramatic play preschoolers enjoy realistic props. But elaborate costumes are not necessary. A cowboy hat or a paper crown adds enough of a theme to launch the active imagination of threes and fours. By five some may want more in the way of costumes with detail. Cowpokes will want a hat, chaps, and even boots, and young royalty will love a cape and scepter. A collection of old clothes, handbags, hats, and beads can be stored in an easily reached drawer or "costume" box or on closet hooks.

As their view of the world and the people in it expands, preschoolers also enjoy playing at grown-up jobs. A collection of hats from Toys To Grow On or Childcraft (see Directory) includes headgear for a fire fighter, astronaut, big-game hunter, and construction worker. From the same mail-order houses comes Grandma's Trunk, a collection of clothes for creative dress-up play that are simple to put on and take off. If you don't have a supply of old clothes (or the time), these ready-made duds may be helpful, although you can probably make your own for much less, and with more fun. A real pocketbook or a velvet cape made from an old dress often has more allure than "toy" costumes.

*Medical Kit*
*Fisher-Price*

## *Props for Play*

Perhaps more important than clothes are the "tools" people use for doing their work. A doctor's or nurse's kit sets the scene for check-ups. A storekeeper needs a toy register and play money. Fisher-Price makes both a sturdy register and a Medical Kit with a working stethoscope.

Both boys and girls love props for pretend journeys. A dashboard with steering wheel, horn, and keys will be used for playing truck driver, train engineer, jet pilot, or computer. The Electronic Supersound Driver by Playskool makes a wonderful toy for make-believe trips when you can't leave home.

Domestic play is popular with both boys and girls. Since homemaking and childcare figure so heavily in their day-to-day life, children are able to step into related role-playing. Toy brooms, dust pans, vacuum cleaners, ironing boards, and no-heat irons continue to be useful props.

Cooking and feeding are favorite themes. A sturdy set of dishes, pots and pans are basic tools for tea parties. The dishes suggested earlier are still appropriate. Fisher-Price has just introduced an extensive line of kitchen supplies, and their "whistling" tea kettle is especially handsome. It sounds off when the handle is lifted. The dishes and pots in this set are also sturdy and attractive. But avoid the fake food. Kids don't need it and they are likely to mouth it. Here's a place where imagination can take over if you encourage it.

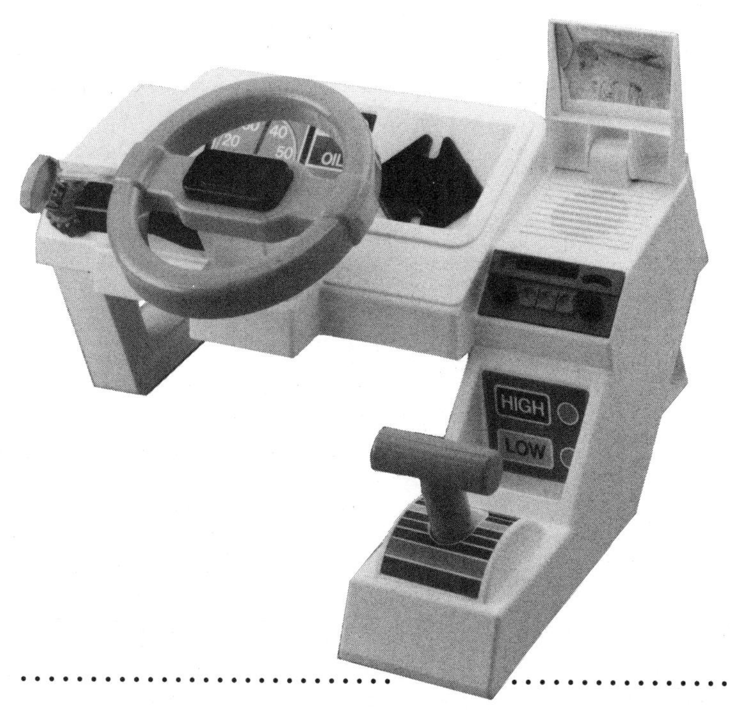


*Electronic Supersound Driver*
*Playskool*

If there's room for a play kitchen, a toy sink with a basin that holds water can be used for washing dishes and doll clothes and for whipping up suds. Little Tikes' Party Kitchen combines oven, sink, refrigerator, table, and chairs into a play center that will be enjoyed for several years. If space is tight, their combination kitchen is more compact but almost as versatile. For even less space and much less money, the Fisher-Price Combination Sink and Stove will fit on any surface and can be packed or stacked away when not in use.

For play-acting there is nothing more appropriate than toy playhouses, tents, or even cardboard boxes decorated to look like a playhouse. They provide privacy and a separate little world the child can step into and take charge of. A big vinyl playhouse from Little Tikes comes with windows and a working door. It can be used indoors or out, and the little Dutch door can be used for a puppet stage.

*Compact Kitchen*
*Little Tikes*

Miniature airports, farms, and garages that come with multiple pieces are often enjoyed for playing out little dramas or for combining with blocks. The Fisher-Price Play People House provides family and furnishings for domestic dramas. This is more appropriate for preschoolers than an elaborate dollhouse, which older children will enjoy. Also interesting for active play is the Fisher-Price garage or airport.

Keep in mind, however, that for children who still put toys in their mouths, plastic and wooden playsets with small multiple pieces can be a choking hazard. Don't depend on the age labels. Depend on your knowledge of your child.

For children who have passed the mouthing stage, a classic Noah's Ark from TC Timber will get years of use. There are thirty-two wooden animals along with Noah's family, and the ark itself provides storage space for all aboard. They'll use all the pieces in this set as props for playing zoo, circus, and farm with their blocks. You'll find it in the Toys to Grow On catalogue.

*Noah's Ark*
*TC Timber*

This is the time for a wooden train set with tracks and bridges that can be assembled in numerous ways. Brio's magnetic trains offer an open-ended variety of accessories that can be added on birthdays and holidays for several years of pleasure.

Preschoolers are fond of baby dolls that drink and wet, have eyes that open and close, hair to brush, and clothes to change. At least one doll should be waterproof for play in the tub. Childcraft makes brother-and-sister dolls that are anatomically correct, and bathable, too. Novelty dolls that talk, creep, and walk generally lose their appeal before long. A soft huggable cloth doll will say whatever your preschooler wants to hear. Some of the loveliest are made by Pauline and Corolle. This is the time when a Raggedy Ann makes a fine bedtime and tea party companion.

Boys as well as girls enjoy playing with baby dolls and soft cloth dolls. The Cabbage Patch Kids and Hasbro's My Buddy doll have filled the void for boys. Other companies have made boy dolls in the past, but they gave up in the face of market resistance. It's only recently that attitudes toward

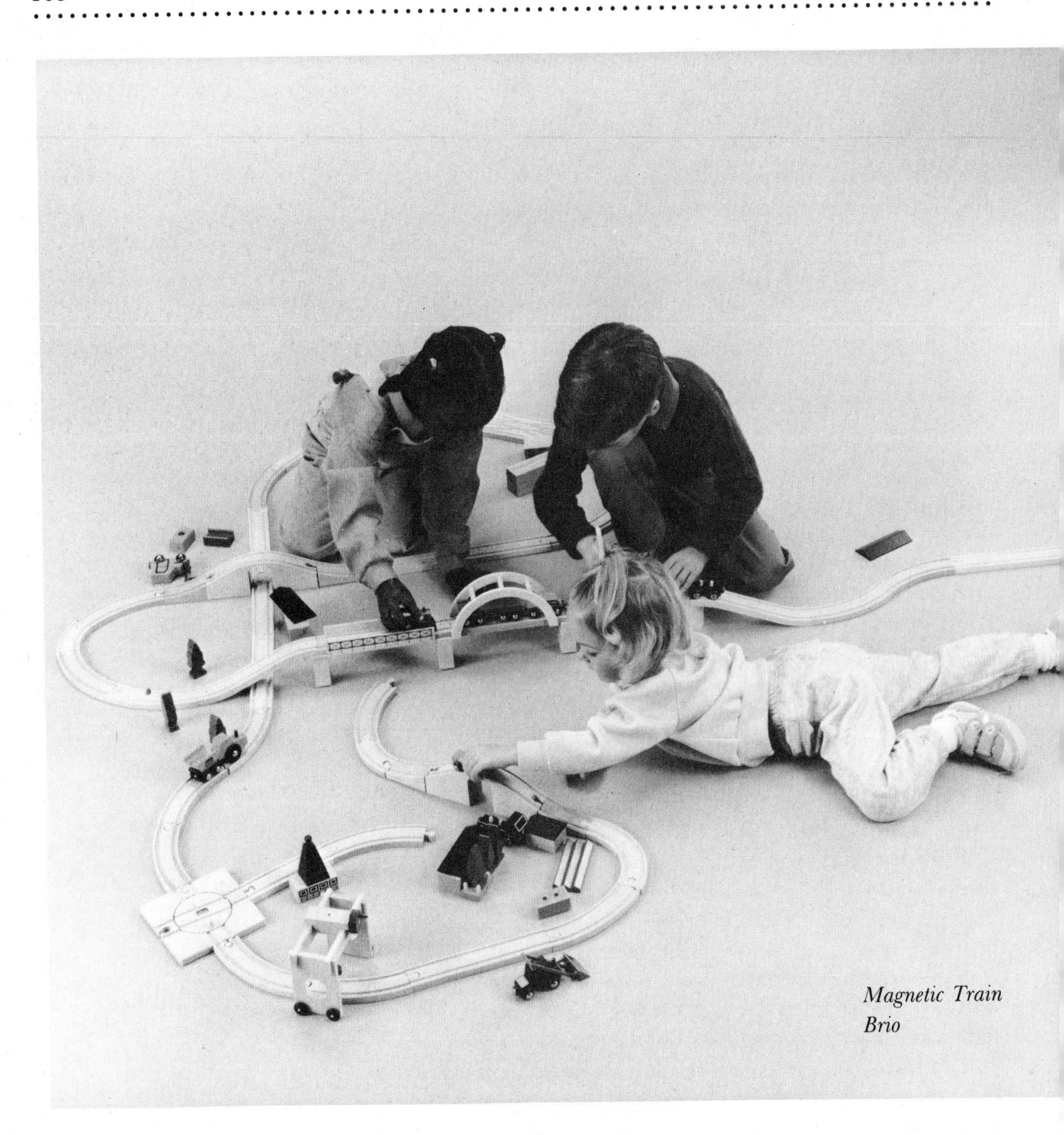

*Magnetic Train*
*Brio*

boys and men have expanded to include nurturing and childcare—at least in some homes.

Increasingly, toymakers are also acknowledging that we live in a multi-ethnic world and that children enjoy dolls that reflect themselves. In the past, the few black dolls available were merely white dolls with tinted skin. Now there is a new generation of realistic and appealing dolls with black, Hispanic, and Asian features. Huggy Bean from Golden Ribbon Playthings is a soft and huggable black doll. Childcraft has both boy and girl dolls with Hispanic and Asian features. Among the most charming boy-and-girl dolls are a Chinese brother and sister from Dolls by Pauline.

As doll play becomes more elaborate, so does the need for doll accessories. A carriage or stroller, a bed or cradle, and a high chair are all useful additions. Badger Basket's Wicker Cradle will accommodate a good-size baby doll or

*Cabbage Patch Kid by Coleco and Wicker Cradle by Badger Basket*

*My Buddy*
*Playskool*

two. And Mattel makes a doll-size Aprica stroller that looks just like the real thing.

Fashion dolls and their clothes are too difficult for preschoolers to manipulate with ease and independence. For that reason, Barbie will be enjoyed more a few years down the road. Right now stick to baby dolls with easy-on and easy-off clothing that fastens with Velcro or easy snaps and buttons.

## *Plush Animals*

Plush animals are also favorite childhood companions. Teddy bears, monkeys, kittens, and other creatures come in all shapes and sizes. Often one special plush toy will take on more importance than others. Like Linus's blanket it is carried to bed and on outings. It is privy to secrets and subject to severe scoldings and hugs and kisses. This very special

*Dinosaurs*
*Manhattan Toy Company*

toy may need to be patched up and accepted, no matter how ratty it becomes. Your efforts to replace it will probably fall on deaf ears. For a preschooler, such a toy offers comfort and a tolerant soundingboard for emotion. In the struggle for independence, the beloved toy offers a dependable source of home and security. And unlike friends and family who have opinions, Teddy has no opinions except those of his maker. No wonder he's such a comfort as a constant companion.

## *Sex-Typed Toys*

For some fours and fives a hook and ladder truck is the gift of gifts. However, one parent who gave her daughter a beautiful fire truck reported that her daughter was offended and repeatedly asked, "Why did you buy me that?" In her eagerness to buy nonstereotyped and nonsexist toys, this mother discovered that her daughter's view of girl toys versus boy toys was more rigid than her own.

Although parents may wish to extend the play possibilities for their sons and daughters, preschoolers are already often very much cued into sorting the world by gender. This should come as no surprise, as one of the developmental tasks of this age is to establish one's own sexual identity. In doing so, the preschooler may have more conservative views than parents would wish of what boys and girls should wear, play with, and do.

On the other hand, many children enjoy dressing up in clothes of the opposite gender and trying on both male and female roles. Through pretend-play, children can safely transform themselves without risk, since they know it's just a game. Their interest in "cross dressing" is usually short-lived and should not be worrisome to parents at this stage.

Studies show that even though women's roles in the family and the work force have been changing, boys still tend to play more rough-and-tumble games than girls do. And

*Sesame Street Puppets*
*Hasbro*

although girls' options in toys have expanded, the same studies indicate that they still prefer dolls and domestic play. But how much of this is cultural? Girls' toy options are expanding, but certainly within limited boundaries only. Take, for example, She-Ra, sister of He-Man: she possesses magical powers but is not a warrior like her brother. Indeed, her canopy bed is a lot closer to Barbie's than it is to something a superhero would sleep in.

For parents who are concerned about raising their children to be nonsexist, remember that preschoolers learn best from the models they see. If they grow up with parents who do not divide the world into women's work and men's work, children will certainly get the message.

## *Puppets*

Through puppets, children can say things they would never say in reality. They can ask questions, give answers, and clarify their feelings. Unlike role-playing in which they transform themselves, in puppet play the child is once removed. Rather than using his own body, the child projects his feelings, ideas, and story on to the puppet through action and words. And with a puppet on each hand, the preschooler can play two roles at once!

Puppets can be used solo, with parents, or with friends. Children often use them to replay familiar stories or to play out stories from their inner lives.

Reasonably inexpensive plush animal puppets are available from Dakin and Cal Toy. For many Sesame Street fans, the familiar Big Bird, Ernie, and Cookie Monster from Hasbro are naturals; they come with premade personas that can be used as a springboard to original stories or to replay meaningful themes from the show.

Useful, too, are semirealistic Poppets puppets, such as a doctor, a nurse, and family players. None of these puppets needs a puppet stage. A table top draped with a cloth can serve as a barrier to hide behind. Or a curtain on a rod can be suspended in a doorway to create a stage.

Homemade puppets made with paper bags, socks, and scraps are another, less expensive, way to expand the cast of players.

## *Transportation Toys*

Preschoolers have a fascination with wheeled toys they can use for dramatic play, indoors and out. Realistic trucks with details and moving parts are most interesting for the sandbox or to use along with blocks. The Mighty Tonka dump truck, car carrier, earth mover, and fire truck are all good choices.

*Mighty Tonka Earthmover*
*Tonka*

Small Hot Wheels and Matchbox toys are also popular for pretend-play. A playmat with roadways adds a mazelike setting for action. Such cars also go well with blocks for dramatic play. As the collection grows, a toy carrier can serve to keep order. Mattel's Cargo Carrier is both a working toy and a storage case that will load up to ten cars.

Although many adults can't figure them out, five-year-olds seem to have no trouble with Transformers. Combining manipulative play with drama, the robots can be transformed from vehicles into warriors. While one could wish this kind of armored toy were not the dominant play theme of the mid-1980s, they are certainly in sync with the preschooler's interest in transformations and fantasy. (Ride-on transportation toys are discussed along with outdoor toys; and wooden trains will appear in the section on block play, below.)

## *Blocks*

Few toys are as ideally suited to the preschooler's playing style as a set of wooden building blocks. While toddlers use blocks as manipulatives to bang, stack, fill, and dump, preschoolers use them as the raw material for more imaginative play. If you've bought blocks and they're not being used, it may be that you haven't provided a sufficient quantity and variety of shapes. The typical store-bought starter set of twenty blocks is hardly enough to get anyone started. Although an adequate set of sixty to seventy blocks with eight or nine shapes is expensive, few playthings will last or hold their interest for as long. Think of blocks as an investment that will outplay almost any other toy you buy.

Unlike a toy garage or farm that comes with a preset structure, blocks can "become" whatever the child wishes. The variety of themes and scenes children construct come from the child's imagination rather than the toymaker's. As with sand, clay, paint, and other unstructured materials, blocks put the child at the center of his play. They invite the child to invest himself in creating his own symbols—his own bridge, tower, airport, or playground. This ability to make one thing stand for another takes the child a step closer to dealing with more abstract symbol systems like words and numbers. Moreover, it is another way of playing with reality

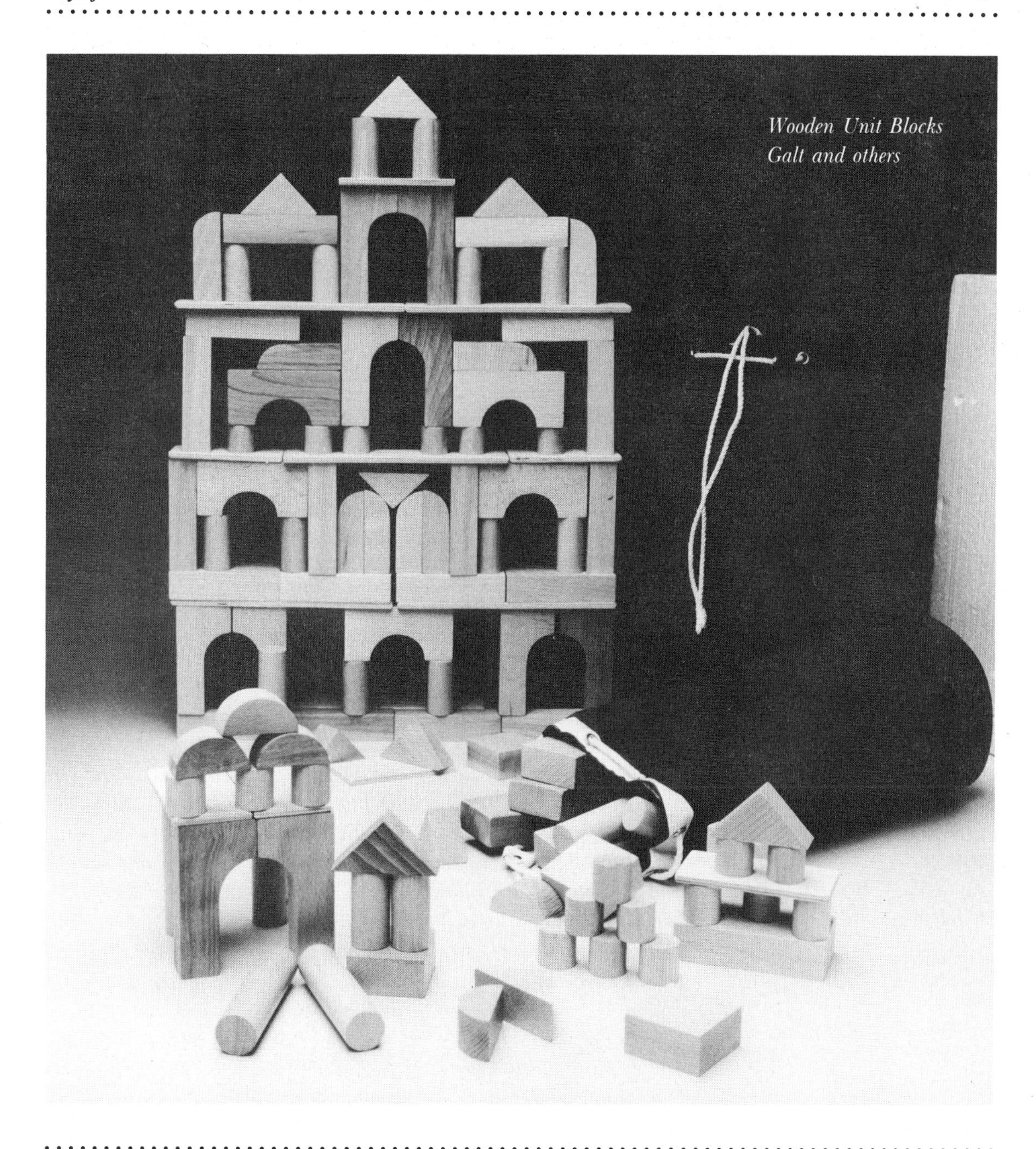

*Wooden Unit Blocks*
*Galt and others*

through fantasy, of bringing the world down to the child-size proportions he can handle.

At three, most of the block builder's energies will be spent in the construction itself. By four and five, building is just the beginning. Now their more complex and elaborate structures become the setting for small dramas. In this way the child set builder also becomes the stage director, actor, and audience.

Whether children are playing alone or with others, block play is rich with many layers of active learning experiences. In playing with others, there's a need for social dialogue and for cooperative planning. Alone or together, block builders learn to solve problems of balancing bridges, ramps, and towers. They learn to accept a bit of frustration in exchange for the pleasure of getting things to "work." Blocks have a way of stimulating both doing and thinking, and they expand the child's attention because the task is self-made. When children knock over their own creations, they're not worried about breaking things that can't be fixed. In fact, they feel a certain sense of power in being able to break down what they have built and start all over again. It's theirs, and that's a satisfying notion, too.

Standard blocks come in multiple sizes and have a logic that children discover and use with amazing ease and skill. Given opportunities to experiment, the young block builder discovers that four squares equal a bigger square, and two triangles make not just a ramp but a rectangle. Consider, too, the mathematical concepts and language young builders directly encounter. Abstract relationships like more and less, shorter and taller, under and over, same and different, wider and thinner, and top and bottom are all less slippery when children use them and experience them concretely in the context of play. Geometric shapes are not flat outlines on a printed page but three-dimensional objects that can be examined, tested, and used for a purpose. The difference between a square and a rectangle is not an exercise in words but a sensory experience that children can hold on to. In playing with blocks, children slowly acquire more than labels for squares, triangles, and rectangles. They begin to grasp meaningful concepts that are the foundations of mathematics. Even without formal lessons, they learn solid lessons from their block structures.

If your budget won't cover the cost of a set of standard blocks, you may have the tools and skills to cut and sand a set from scrap lumber. Make rectangles twice the length and oblongs three times the length of the square units. So if your basic square is 3″ × 3″ × 3″ your rectangle will be 3″ × 3″ × 6″. Still less costly is a set of blocks made from milk and cream containers. You'll need to cut and tape the ends of the half-gallon, quart, pint, and half-pint containers in order to square them off. And although they lack the heft and variety of wooden blocks, they are an inexpensive alternative.

If space is an issue, another alternative that's also less costly is small-scale blocks. A sixty-piece set in this size will cost much less than a full-sized set but offer the young builder quantity and some variety of shapes.

## *Props for Blocks*

Although the older preschooler can take the flight of imagination that transforms a block into a train, a tower, or a couch, block play is definitely enhanced with a supply of realistic props. Small vehicles, people, animals, and dollhouse furniture should be stored near the blocks. If you're buying small-scale blocks you'll want to provide small-scale props that fit.

Older fours and fives take pride in telling you all about their constructions and how they work. They are sometimes reluctant to knock them down and may want to play with them over a period of several days. Fives, who are becoming more interested in signs, may appreciate your printing a

*Mega Blocks*
*Ritvik*

sign for their Empire State Building or their Golden Gate Bridge. Taking the time to admire their work and taping a sign here and there lends support to their play. It's also a way of connecting kids to the world of words and print without formal lessons. Labels are often the first "sight" words pre-readers learn to recognize. Naming and labeling their creations help them understand that the words we say can be written down. This is something we take for granted, but children need to learn it in ways that are personally meaningful.

Keep in mind that the way blocks and props are stored can affect the quality of other play. A storage box is less desirable than an open shelf where blocks can be sorted and stored by size. Blocks and toy vehicles, people, animals, and furniture can be stored on shelves and in separate baskets or shoe boxes rather than thrown together in a mixed-up jumble that has to be dumped in order to find a particular item.

Providing this kind of order makes putting toys and blocks away into a game that encourages thinking and classifying rather than disorderly digging and dumping. Although preschoolers often need help picking up, you can get extra learning in if you say, "I'll find all the animals, you pick up all the trucks and cars," or "Let's find all the longest blocks first."

Blocks are enjoyed by girls and boys alike. Girls often build houses, playgrounds, schools, and farms. Boys are more likely to build airports, cars, and forts. He-Man and his action cohorts will certainly find their place in adventure-filled block dramas. Indeed, blocks are an ideal setting for working through some of the confusion and feelings children have surrounding cartoon characters. Here they are in charge. With blocks they can also discover they don't need all the accessories pressed from a plastic mold. They can create them themselves from their own constructions.

## *Construction Toys*

Many of the building sets designed for preschoolers offer a combined challenge to imagination and dexterity. Legos,

*Imaginetics Magnetic Blocks*
*Skilcraft*

*Warning: Toys with parts small enough to swallow or choke a child are usually labeled "Not recommended for children under three," on the assumption that threes have passed the stage of putting things in their mouths. However, many threes, fours, and even fives do put things in their mouths, ears, and even noses. If your child still does so, you'll need to avoid toys with small pieces—no matter what the label says!*

Tinker Toys, Mega Blocks, and Skilcraft's Imaginetics magnetic blocks all work very differently and produce different results. The small bits and pieces require less space than blocks and may be easier to cope with in small apartments. However, small pieces may be a choking hazard to younger siblings and should be stored out of reach.

Although most construction kits come with illustrations of end products, don't get stuck on building a model of what's on the box. Give children plenty of time to experiment with the pieces and invent their own designs. Reproducing a predesigned model is a worthwhile skill, but it is more easily accomplished after a period of free play and should not rob the child of the chance to experiment on her own.

Studies with preschoolers show that children are better problem solvers after they have time to explore materials than if someone demonstrates what to do and how to do it. Be there to help if they ask, but don't take over and build or direct the action step by step. Mistakes are a way of learning, too, so leave room for some trial and error.

You will probably find that some of the small and less expensive starter sets are too limiting and make better add-ons than introductions. A beginning Lego set will get more mileage if it comes with a platform and wheels so that "working" toys can be created right away.

## Art Materials

Unlike the toddler who enjoys scribbling with crayons, pinching and poking clay, and generally manipulating art materials, preschoolers grow increasingly interested in the outcome of their actions. The same kind of symbolic play that colors their construction and dramatic play is expressed with art materials. With their expanding control of hand and eye, their paintings and drawings become less accidental and more purposeful. With pencils and crayons they are likely to draw representations with greater detail and control. Even so, the end product should not be the only goal.

Again, unstructured materials invite the child to invent and create his own structure and story. Learning to color inside of lines comes more easily after children have experience with the more exploratory business of making their own lines and filling in their own shapes. It is not until they are five that children enjoy the structure of coloring inside the lines as well as drawing their own pictures.

Rather than coloring books, provide preschoolers with pads of paper, large sheets of newsprint, wrapping paper,

*Safety Scissors*
*Binney and Smith*

and colored construction paper. Scrap paper from a printing plant or old wallpaper samples from the local decorator are all good sources of inexpensive and plentiful supplies. Also wonderful for drawing are rolls of shelf paper that can be used like a scroll, and computer printouts that might otherwise end up in the wastebasket.

Preschoolers are also ready for trying their hands with scissors and glue. Binney and Smith scissors work for righties or lefties, cut a clean line, and are safe enough for children to handle independently. White liquid glue like Elmers is easier to handle than most. Scotch Tape also offers new ways of putting materials together.

Crayons should be big and chunky. They're easier to grasp and are less likely to snap than the standard-size crayons that schoolage children use. Pencils wide and narrow, watercolor markers, and chalk all produce very different results, providing delicious variety for eyes and hand.

Fours and fives often want to know how to write their names. To help them learn how, simply print the child's first name, using a capital for the first letter and lowercase letters for the rest. Parents often print the whole name in caps because they think big letters are easier, but it's just as easy for children to learn it the right way from the start. Whatever happens, don't get fooled into thinking this early interest is a signal to teach penmanship. Children who play with writing tools are getting ready without formal lessons. By all means show them how to write if they ask and tell them what other words mean when they inquire, but don't push. Too much drill and practice can be tedious and may sour attitudes about school-related skills.

More appropriate for now are the free-flowing colors that a brush and tempera paints can provide. An easel isn't necessary. In fact, some experts believe that a flat table makes it easier to control the paint—there are certainly less problems with running paint. On the other hand, having an easel

OPPOSITE
*Easel*
*Little Tikes*

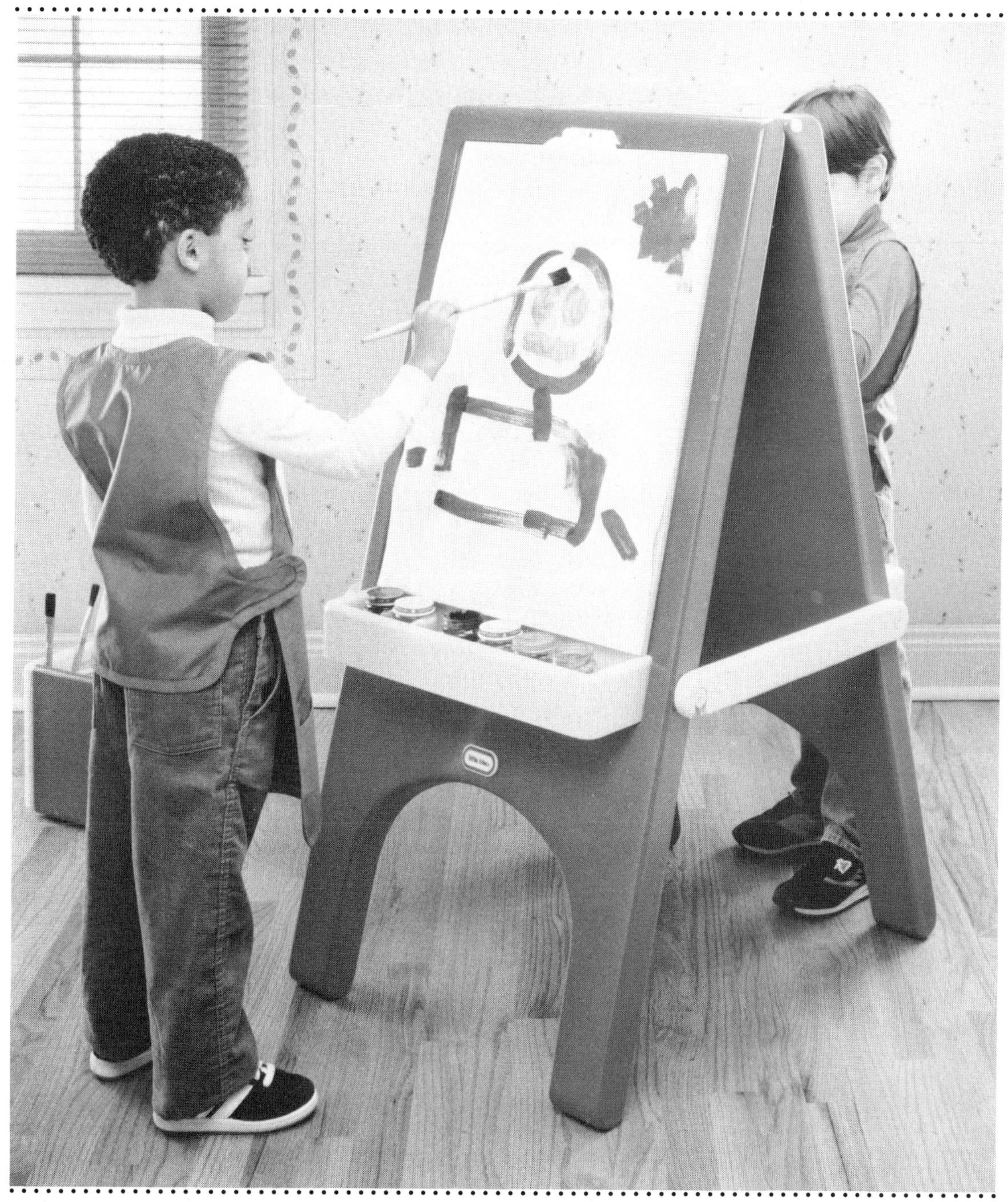

set up makes paints or chalk accessible without much assistance. One doesn't have to clear a table, get set up, and put everything away each time. As a result, painting may be more of an everyday activity than an occasional event. Little Tikes' easel is solidly built and easy to scrub off. One surface can be used as a chalkboard.

For preschoolers, liquid tempera paint (poster paint) is more desirable than cakes of tempera or watercolors. They're brighter and more opaque, so wobbly strokes are more defined. Long, stout-handled brushes ½″ or ¾″ wide are a better choice than tiny watercolor brushes that are hard to control or get much coverage with. Buy the three primary colors (red, yellow, blue) to start with, and add white and black later. An art-supply shop or school-supply company will have larger jars at a lower cost than most toy shops.

Of course, you won't want to put the whole paint supply out at one time. Preschoolers are apt to plunge a blue brush into a yellow paint pot. To keep paint clean and keep it from drying out, pour small amounts into covered containers. Galt makes a set that will fit on the easel; you'll find them in Childcraft and other toy catalogues. Also useful for paints, chalk, crayons, glue, and markers is Rubbermaid's utility carrier, a plastic tote with a sturdy handle. It works well for storing supplies, and the child can have the art area where the project is. You'll find these in housewares departments. They're useful for storing construction toys, too.

Another option for beginning painters is a set of Squish Paints from Adica Pongo. The self-sealing paint containers sit firmly on tabletop or easel and are spillproof. Or, for more instant paint-and-brush action, look for Magic Brushes, a set of squeezable tubes filled with paint that flows onto the brush tip. Compared with a set of paints, these are costly and limited. The brush is narrow, and it's difficult to mix colors. But it takes the hassle out of setting up paints and worrying about spills, and they're totally portable for young landscape artists.

Big colored markers are also wonderful for small hands. You'll want to be sure the markers are water-based and washable. Of course, preschoolers are apt to forget to cap them, and they do dry out. Most also tend to bleed through paper and will mark up a tabletop, so be forewarned. Once the necessary precautions are taken, though, they give dandy results.

## *Finger Paints*

Preschoolers also find finger paints especially satisfying. Here's a legitimate way for them to get up to their elbows in colors! There's no brush to get in the way and the action involves big muscles as well as fingertips. For a lift on a rainy day, there's probably no toy they'll love more.

You can buy finger paints or mix up your own (see recipe in chapter 4). Either way you'll need slick shelving paper for your children to work on and a good flat surface. Childcraft's Messy Play and Hobby Tray can be washed off and stored, and can also be used for clay. A plastic-topped table will also do the job and is easily cleaned. Involve your child in the business of cleaning up after using art materials. For kids, washing the brushes and the table is part of the fun and helps them learn about responsibility.

Your preschooler may enjoy telling you stories about the painting she makes. Listen if she volunteers, but don't insist that she show and tell about each and every painting. With older children you may want to write down the story they dictate. Seeing their words turned into letters and words is another step toward reading. It helps them understand that words we say can become words we read.

If your child needs a bit of encouragement to start talking, offer comments about the color and shapes she has used, rather than asking "What's that?" and pressuring the child to make representational paintings. For instance, you might say, "I like the way you used blue dots" or "That's a big red circle you made." Often your comments will oil the wheels of conversation and lead to lengthy tales.

Although preschoolers are not terribly concerned with end products, they do like to have their artwork displayed on the refrigerator or on the wall. An inexpensive box frame that lifts open can be used and reused for a changing display that says indirectly that you value their creations.

## *Clay and Modeling Materials*

Play-Doh and homemade dough continue to be of interest for modeling and manipulating. Play-dough shaping kits with preformed presses are more mechanical than enriching. True, children enjoy making recognizable shapes, but too many such toys take the playful unstructured aspect of art material out of the child's hands. A set of cookie cutters, jar lids, and a dull knife will work the same kind of "magic" without encouraging the child to rely on tools that rob her of play. When the end products are too perfect, the child may come to devalue her own rougher creations. Kenner did introduce a few new Play-Doh Gadgets this year that are more open-ended and will give Play-Doh some new twists.

More interesting, if also messier, is real clay. Self-hardening clay may be bought in an art-supply store or by mail from school-supply houses. It's sold by the pound and must be kept moist in a plastic bag. A one-pound box of self-hardening clay will go a long way. It offers small hands an altogether different experience than dough, which tends to be less responsive to the touch. Preschoolers also love to experiment with clay, adding water and enjoying its slippery feel. Although preschoolers like to do and undo their modeling creations, older fours and fives like to name the things they have made and often enjoy keeping them and even painting them.

Another modeling material commonly found in toy stores is Plasticine. This is a nonhardening material that comes in bright colors. Big blocks of Plasticine are extremely tough for small hands to manipulate, but the material softens to the warmth of hands and holds its shape well. Before handing over a wad to your preschooler, take the time to soften it up.

## *Collages and Constructions*

You'll find ready-made kits in arts and craft shops for collages and constructions, but you can also begin collecting a scrap box of materials that can be glued, taped, or wired together. Most of the materials are in anyone's house. What's not there now can be added as children become involved in spotting "good stuff." Here's a partial list to start with:

- toothpicks
- pipe cleaners
- straws
- shells
- foil
- shredded paper
- bumpy cardboard
- tissue paper
- bits of fabric
- cotton balls
- styrofoam packing material
- cellophane

All of these multitextured materials lend themselves to cutting, pasting, and arranging in original designs.

For less messy (but often less satisfying) design experiences, there are plastic Colorforms as well as Childcraft's Magnetic Picture Blocks with geometric shapes that can be used in the same fashion and stored in their own carrying case.

Such toys may be especially useful if you are traveling and don't want to deal with clutter and mess. Also satisfying for

*Magna-Doodle*
*View-Master/Ideal*

travel and for convenience at home is the classic Etch-a-Sketch from Ohio Art. Twirl the dials and draw a design. Or for more direct control, there's Magna-Doodle from Ideal. This toy works like a chalkboard so a child isn't turning dials but drawing directly on the board. And don't forget the perennial favorite, the Magic Slate, which is more direct and therefore provides somewhat more control. These old-fashioned cardboard pads with acetate sheets you write on and lift to erase are easy to take along. They're inexpensive and don't last long, but they're convenient and they satisfy the preschooler's love of scribbling.

Since storage of art materials is often a hassle, look for lap desks that provide a flat working surface and compartments for drawing tools. These travel easily on long car trips or from room to room. Childcraft has a clever chalkboard lid that slides over a storage box, perfect for holding chalk, eraser, and other toys. For floor sitting or a day in bed, a Play Table, also from Childcraft, provides a surface for chalk that wipes clean and recessed compartments for paper, crayon, scissors, and other supplies.

When it comes to art supplies, don't be seduced by the complicated kits in the toystores. Too many of these kits offer less than meets the eye. Heavily structured, with precut stencils and grids, they take the art right out of the child's hands. Children who are encouraged to use a variety of materials will experience the excitement of making something out of nothing. By allowing them to solve their own problems and use their own imagination, such art experiences are rich with learning that goes beyond the confines of a stencil or a coloring book.

## *Music and Motion*

For preschoolers, music is more related to motion and physical activity than to quiet listening pleasure. The rhythmic pulse of a march, waltz, polka, or hard rock will set them in motion. Quite naturally they respond to a change of

*Tape Recorder*
*Fisher-Price*

tempo with body movements that are remarkably syncronized. And with homemade or store-bought rhythm instruments, they can add to the beat. Preschoolers love to dance, sing, and play along with all sorts of music. As in all their play, music lends itself to dramatic interpretation and to stories that can be played out with the whole body or used to inspire drawings and paintings, playful lyrics and fanciful daydreams.

An easily operated phonograph or tape player puts children in charge of their music and gives them the pleasure of pushing buttons and operating their own equipment. Fisher-Price's Tape Recorder and Phonograph are both ruggedly built and simple to operate.

Build a tape or record library with plenty of variety. Folk songs, marches, light classics, as well as Latin and other ethnic music are all exciting for children—some more exciting than music packaged especially for them.

If your child wants to make music of her own, take a trip to a music shop or check a school-supply house for rhythm instruments. Few of the toy instruments found in toystores have good sound quality, and most are quite flimsy. A solid drum or tamborine, or a set of cymbals, maracas, or wrist bells will be enjoyed for several years.

Good for tooting along with tunes is the Fisher-Price Crazy Combo, a horn that fits together in multiple configurations. Toy xylophones and pianos are also fun for picking out tunes. Older fours and fives will enjoy following the color-coded music that accompanies Little Tikes' Tap-a-Tune, suggested earlier. However, for true musical quality you may want to consider investing in the Chimalong. You'll find details in the following chapter.

For quieter moments preschoolers also enjoy hand-cranked music boxes. Although the sound quality leaves much to be desired, a hand-cranked guitar is often a favorite for dramatic play. It can also be taken along when you travel.

## *Toys for Active Play, Indoors and Out*

Running, jumping, climbing, swinging, chasing a ball, or walking a balance beam—what whirlwinds of activity preschoolers are! Why? Because they have a genuine *need* for active play. Even if they're watching television or listening to a story, the preschooler rarely sits still voluntarily for long. Whether it's hot or cold, rainy or fair, young children need some time each day to stretch and flex their muscles and to test their own strengths and limitations. It's through their bodies that children first gain an inner sense of themselves as able and competent doers. "Watch this," the preschooler calls as she climbs the slide or hangs from the monkey bars. Unlike adults, children don't separate physical activities from thinking and feeling. Active play is no less important to their sense of well-being than other forms of play. It not only serves as an outlet for tension, but it also stimulates their imagination and intellectual development.

In their desire to test limits, their own and others, preschoolers are apt to take risks. As a result, they still need supervision. By five, most are better able to predict how far up they can climb or how far down they dare jump. Even so, they can get carried away with their own exuberance. He may climb to the top of the monkey bars and then lose the courage to come down. His own energy and enthusiasm may lead him into some tight spots, but for the most part, the preschooler's well-developed gross motor skills are a source of pride and give the child a sense of accomplishment.

In the backyard or park, there is room to hoot and holler, run and jump, laugh and scream without being hushed or scolded. There are no downstairs neighbors to disturb, no vases to break, no babies to wake. Active physical play offers the young child an appropriate outlet for his seemingly endless supply of energy. It also puts him in situations where he's likely to meet and make friends.

## *Pedal Toys*

Threes are ready for their first pedal toys. For threes, fours and most fives, a three-wheeled trike is more appropriate than a two-wheeler with training wheels. For the child's safety and sense of competence a ride-on toy needs to be the right size. Don't look for something they'll grow into. Getting on and off should be easy and their feet should be able to touch the ground when they sit since there are no brakes. A small child may need a trike with a ten-inch front wheel although the average trikes are between twelve and fourteen inches.

If you're buying a first trike and planning to pass it on to younger family members, look for a classic red trike rather than one that's plastered with licensed characters. After all, your infant son or nephew won't want his big sister's pink or Cabbage Patch trike, even if it *is* "just like new"!

Lightweight plastic trikes have some advantages over the traditional metal ride-ons. The bodies and handle bars don't rust, although the mechanisms still do. For city-dwellers an easy-to-lift trike is a real plus for getting in and out of elevators and managing sidewalk crossings. Low-slung three-wheelers also tend to be less tippable. However, they also represent a greater hazard in that they are low to the ground and harder for cars to spot. Of course, preschoolers should not be riding in the path of any vehicles. But the difference between "should" and "could" can only be handled by supervision. Be sure to have your child test-drive before you buy. Some of the low plastic trikes generally recommended for fours are impossible to adjust for height, so they don't fit for long.

Some fives are able to ride two-wheelers with training wheels. If they have older siblings or if two-wheelers are the "in" vehicle in the neighborhood, you may want to consider it. Again, safety should be the first consideration. If your five-year-old is very small, a two-wheeler with training wheels may be out of the question. Even if the bike has brakes, he should be able to straddle the bike and put his feet flat on the ground. If not, getting off and on represents a potential hazard. Although toward the end of the year *some* fives do learn to ride without training wheels, most will benefit more from the security and ease of a solid three-wheel trike or the steadiness that training wheels provide. Occasionally your child may bite off more than he can chew. If he has a two-wheeler with training wheels, he may insist he can ride it without the training wheels—and then may end up needing them again! You be the judge.

Fours and fives also enjoy highly realistic and detailed toy cars, fire engines, and tractors with horns, steering wheels, and plenty of potential for dramatic play. Unfortunately, most of the new models operate on battery power instead of pedals. There's no doubt that children love the magic of the mechanical models, but these more expensive battery-powered ride-ons have less to offer in terms of physical development.

In keeping with their interest in action and pretend-play, threes and fours will find multiple uses for small wagons to pull and wheelbarrows to push. Other wheeled toys that resemble grown-up tools are also useful. A Fisher-Price lawnmower that just happens to leave a trail of bubbles is a favorite of threes and fours. A stroller or carriage for taking a baby for walks is useful indoors and out. Child-size rakes and snow shovels are also inviting tools for active role-playing out of doors.

Older preschoolers also like to test their balance in new ways. A pair of double-bladed ice skates or plastic-wheeled

Fisher-Price

roller skates are a challenge for older fours and fives. Fisher-Price's Roller Skates are adjustable for growing feet and can be locked into a forward-only position for beginners.

OPPOSITE
*Lawnmower*
*Fisher-Price*

By five, most children have a solidly developed sense of their own bodies. Their gross motor coordination is secure and fluid. They can not only run, jump, and climb, but are also learning to skip and hop. Indoors or out, a jump rope is a playful challenge. They're fascinated with acrobatic tricks and love the thrill of balance beams, somersaults, and tumbling. For indoor use, a large mat from a sporting goods store will cushion their rough-and-tumble play and encourage them to use their whole bodies. If you have room for it, a small trampoline saves bed and sofa springs and provides a legitimate place for active indoor play. Five is also a good time to introduce beginner's skis and a simple sled.

## *Gym Equipment*

If you have a backyard these are the years your children will enjoy climbing equipment, slides, and swings. Although wooden gym equipment is often more costly than painted metal gym sets, it's also more attractive and less "cutesy." Wooden gyms also offer climbing equipment rather than just swings and slides. The problem with highly decorated swing sets is that they are easily outgrown and have limited appeal. If you have to choose between a climber or a swing, slide, and see-saw, go for the climber. You can add a board with a cleat to turn any climber into a low-to-the-ground slide. It also provides a slanted surface to walk on. A tire hung from a tree will make a fine swing, and a see-saw is only useful if you've got two children about the same size. On the other hand, a solid climber lends itself to active physical exercise as well as dramatic play.

If you are buying a gym set, look for swings with curved, soft (not rigid) seats. They're easier to get on independently and won't hurt as much if they whack someone. Slides should have side rails and all equipment should be checked for rough edges and protruding nuts and bolts that could cut or catch small fingers. Depending on where you're setting up the gym you'll need to check the manufacturer's recommendation about securing the equipment firmly into the ground.

Creative Playthings offers several wooden gym sets including an assemble-it-yourself kit that is considerably cheaper than ready-made.

*Activity Gym*
*Little Tikes*

For a simpler and less costly but satisfying climber consider the Activity Gym from Little Tikes. It has slats to

climb, holes to crawl through, a platform to stand on, and a slide.

Other interesting multiple-play structures are Quadro and Playskool's Pipeworks, building sets that can be turned into ride-ons and climbing devices. Parents will need to construct them at this stage.

Of course physical play is not limited to the outdoors. Fours and up will enjoy a Doorway Gym Bar for chinning and other tricks. The sturdy chrome-plated steel bar from Childcraft will extend to fit doorways up to forty inches wide and can be placed at any height.

An inflatable punching bag, or bop bag, is ideal for spending excess energy or unloading angry feelings. The perfect outlet for the wild behavior of fours, it offers a legitimate way to land a punch without landing in trouble.

Another route to make-believe and physical play is a full-size rocking horse with spring action. Though many of the wooden horses are beautifully crafted, some are more decorative than safe. Having lived with three bucking horses, I tend to favor the spring action and soft edges of the vinyl trotters. Personally, I could do without the electronic trotting and whinnying sound effects, but preschoolers love the added touch on their way to imaginary adventure.

## *Balls*

Balls for bouncing, throwing, catching, and chasing continue to be basic equipment. A foam filled Hug-a-Planet makes a pleasing indoor toy and will be a satisfying first globe during the early school years. Nerf balls from Parker Brothers are softer and therefore gentler—both on kids and on furniture. They come in two sizes and many colors. Also fun for bouncing on is Hedstrom's Hoppety-Hop Ball. Fours and fives with their love of role-playing and realism prefer a football or soccer ball that looks like the real thing.

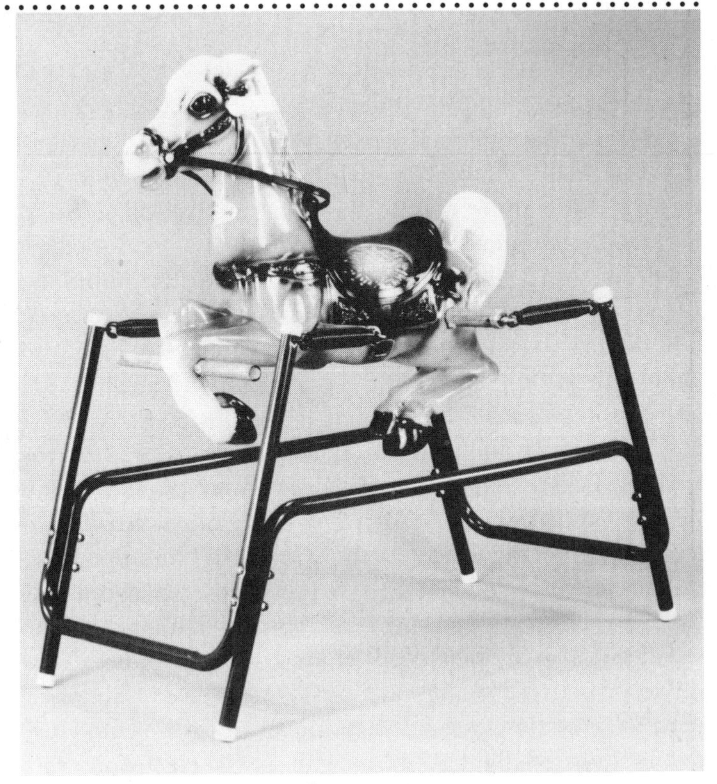

*Clippety Clop*
*Wonderline*

*Hoppety Hop Ball*
*Hedstrom*

Several manufacturers make replicas that are lightweight but realistic looking. However, most are not suggested until five, since they're not as light as a Nerf ball. For young baseball enthusiasts, the soft, fat Wacky Bat and Ball from Childcraft will make a safe hit.

Don't look for much skill or interest in the rules of such games. At this age the point is to act out the game and to have a good time, not to develop specific physical skills.

## Water Play

For the beach or tub, look for boats with action. The Wind-Up Sea Plane from Playskool is easy enough for children to operate independently. Also interesting is a Glug-

Glug boat from Childcraft that runs on water power. You pour water through the top and the boat skims along.

Many of the water toys suggested earlier continue to be of interest. Preschoolers also love the magic of toys that change color in warm water.

For outdoor summer fun, a wading pool is a great place to sail boats, splash around, and cool off. Preschoolers also love the splendid fun of running under a garden sprinkler.

It's a good idea to round up a supply of "tools" for them to use in the bathtub, wading pool, or wherever water play is permitted. Here's a list of common household items:

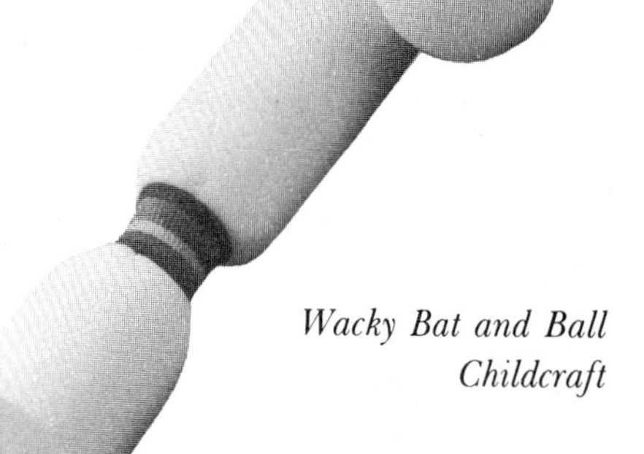

*Wacky Bat and Ball*
*Childcraft*

- funnel
- plastic tubing
- plastic containers
- sieve
- squeeze bottles
- plastic basting tool

## *Sandboxes and Sand Toys*

An ideal sandbox has a flat surface for sitting on or making mud pies. It should be deep enough to "dig a hole to China," and big enough to accommodate two or more children. Most such boxes need to be built and require a backyard. For apartment dwellers or those who don't have the tools and skills necessary, a ready-made plastic sandbox, though it has limitations, is better than none. Little Tikes' new Big Sandbox has built-in seats at each corner, and the cover can be used as a waterway for boats, so it's really two toy in one.

In addition to pail, shovel, and sieve, older preschoolers will enjoy dump trucks, tractors, earthmovers, and cement mixers that actually turn sand and water into "cement."

In the sandbox or at the beach, older preschoolers will also be fascinated by Childcraft's Stacking Sand-and-Water Wheels. They can shovel sand (or pour water) into the funnel on top and watch it spill down and set the multiple wheels spinning. This is a more complex toy than the Sand

*Wind-Up Sea Plane*
*Playskool*

*Sand Castle Play Set*
*International Playthings*

*Big Sandbox*
*Little Tikes*

Wheel suggested earlier. Interesting too are Berchet's See-Through Buckets with clear plastic bottoms that work like viewers. You can look underwater without getting wet.

Fives will also enjoy creating shells, fish, and even castles from molds. Discovering just how much water to add (so that castles don't crumble or melt into mud) calls for patient experimentation—and so does learning how to turn the mold over without losing it. Pails with turret-shaped bottoms and castle-casting sets are wonderful for building castles. Smaller cups and containers come in handy when kids want to shape smaller towers. If you're lucky they'll let you help. And don't put those molds away when winter comes. How about a snow castle?

## *Furniture for Play*

As we saw in the previous chapter, perhaps the most basic pieces of play equipment are child-size tables and chairs. Useful for pretend teaparties and for real mealtime, it's also ideal for puzzles, construction toys, clay, paint, and drawing. Since the table top is going to serve so many functions, be sure that its surface is scrubbable and that the chairs are sturdy. A colorful table from ABI has a flip top that turns into a chalkboard with room to store paper and crayons below. A matching bookshelf provides accessible storage for toys.

A child-size rocking chair lends itself to a little quiet time with motion. For active preschoolers, the soothing rocking action may be useful for looking at a picture book or just plain daydreaming. Some children also like to incorporate rockers into their dramatic play.

Toy storage often becomes a problem during the preschool years. When you purchase a new toy, you can usually tell right away if the box it comes in will hold up. If it won't, toss it out and use something else. The classic toy chest is not only the least satisfactory way to keep toys organized, it can also be a hazard. Old-fashioned toy chests with heavy lids have trapped children's hands and heads. If you have an old toy chest, remove the lid! (See the warning on page 138.) New models have safety devices and lift-off lids, but they invite children simply to throw everything in. More useful at this stage are a variety of containers that can be stored on open shelves or in the closet. Plastic carry-alls are convenient for toting toys from room to room. The type of shoe bags sold for closets are useful for stuffed toys or art supplies. Baskets and stackable cubes are also decorative and functional, and they encourage children to sort and classify their toys. Investing a little time and thought in toy storage will simplify putting things away and cut down on the cries of "I can't find my toy!"

Childcraft has addressed the problem of toy storage in an especially interesting way. Their Play Boxes, ideal for blocks and other bulky toys, can be hooked together like a train and moved with ease.

*Play Boxes*
*Childcraft*

## *Sit-Down Toys*

Active preschoolers also need some sit-down toys for a change of pace. Most of the toys and games in this section require an adult player. Indeed, much of their attraction for the child may be the one-on-one time with parents. An older sibling may be able to fill in, but don't expect a pair of four-year-olds to stick with a sit-down game or puzzle for long.

## *Puzzles*

Puzzles with frames are easier to handle than open-ended jigsaw puzzles. Three-year-olds who have no previous experience will need to start with simple five- and six-piece puzzles. By the age of four, some can handle twenty to thirty pieces. Wooden or foam rubber puzzles are more expensive than cardboard but tend to be easier for small hands to manipulate.

Working with puzzles helps children develop eye-hand coordination and strengthens their understanding of spatial relationships. It also sharpens their ability to fit together multiple parts into a cohesive whole—a skill they will need in reading.

Completing a puzzle also helps children learn to stick to a task and cope with frustration. It can be a source of pride that the child can reexperience with repeat performances. Unlike adults, a child enjoys solving the same puzzle many times before the challenge and pleasure wear thin. When she's ready for a more complex puzzle, the new one should have two or three more pieces. Puzzles with letters, numerals, and pictures are good choices. Also interesting are puzzle pieces that lift to reveal interior pictures, lending the added dimension of a "story" line.

Sewing cards are another kind of puzzle that offers small hands the opportunity to master fine motor skills. Choose rubber cards if you can; they're sturdier than cardboard. A good choice is Galt's Lace Shapes, which are punched out along the outline. Easier to grasp than square cards, their cut-out forms may also be used for dramatic play. Pre-

schoolers may need a demonstration in the art of stitching, so be ready to help. This is another toy that will be used and reused—it's not a one-shot experience.

*Lace Shapes*
*Galt*

Also valuable for refining eye-hand coordination is a set of large beads or shapes to string. Galt's brightly colored

Cotton Reels come with a string with a stiffened end for ease of handling. They can be stacked, rolled, sorted by color, and worn as jewelry.

## *Board Games*

Most board games for preschoolers are of limited interest. It's hard for them to learn the business of taking turns, following rules, and accepting the inevitable fact that sometimes you win and sometimes you don't. Fortunately, there are several games that are easy enough to satisfy them, especially if you're not too rigid about the rules. Classics such as Candyland and Cootie require no particular strategy, and simple games of chance put both young and old players on an equal footing. Other appropriate games allow preschool players to practice matching, counting, naming, and classifying. Look for high-quality graphics in these essentially pictorial games. Those made by Nathan and Ravensburger are especially clear and handsome to look at. Here are some of the best choices for preschoolers:

- Candyland or Match a Balloon: A first board game—teaches color-matching and naming, as well as taking turns.
- Cootie: A first counting game.
- Picture Dominos: A simple matching game—a variation on regulation dominoes with dots, which may be enjoyed by five-year-olds.
- Lotto: Picture-matching games that come in varying degrees of difficulty and differing themes.
- Memory: Like TV's "Concentration" game—players must find matching pairs of cards that are placed face down.

*Lottino*
*Ravensburger/International Playthings*

## *Manipulatives*

Following patterns or creating their own designs helps older preschoolers hone matching skills and other kinds of eye-hand coordination. Wooden parquetry blocks are challenging and more open-ended than most puzzles. These geometric shapes can be transformed into original designs or used with patterned cards to sharpen visual perception and develop a sense of design. You'll find these in most school catalogues.

A game like Peg Pan from TC Timber offers a different kind of challenge to small hands and eyes: children create or copy pictures and designs. To fit parts together to form a whole requires an orderly approach, and children may need simple counting skills and some careful strategies for working one line at a time. Threes and fours may prefer to make their own pictures, while fives are more likely to try

*Kaleidoscope*
*Galt*

to duplicate the preset patterns in addition to making their own.

For threes and fours, Brio's Color Peg Board or their Sorting Blocks challenge the child to explore size relationships, colors, and patterns. So do nesting cups, eggs, and dolls. For a different kind of visual exploration, consider Brio's Cog Labyrinth, an assortment of cogs that can be arranged in multiple ways.

Also interesting for quiet gazing are kaleidoscopes that give changing visual effects. A kaleidoscope with changeable pieces will provide more variety. The kaleidoscope from Galt comes with two colorful end pieces and an empty one that can be filled with various found materials. A marble, a paper clip, a pebble all take on a new dimensions when viewed this way.

For indoors or out, an unbreakable magnifying glass gives young nature lovers and would-be detectives a close-up look at bugs, plants, and paw prints.

Childcraft's giant magnet has a magical appeal as it lifts paper clips, toy cars, and other metal objects. Armed with a magnet and a strip of price-tag stickers from the stationery store, fives will enjoy playing metal detector and labeling magnetic objects with a sticker.

Magnetic letters and numerals or the soft and stickable vinyl symbols from Soft Touch that adhere to mirrors, windows, and enameled walls are interesting for manipulating and naming. Handling the cut-out symbols gives small hands a feel for the shapes of letters, which they can carry into writing and reading. But don't push. Matching letters, spelling their names, or putting numbers in sequence comes after children have a chance to play freely with the various pieces. Remember that early achievement with abstract symbols is not the ticket for success in school or life.

## *Books, Tapes, and Talking Toys*

If you're really interested in building your child's prereading skills, take time to sit down and share the pleasures of picture books. Build a library of books that your child can enjoy with you and look at independently. Don't overlook the possibilities waiting at your local library. Many have regular story hours with audience participation and regularly scheduled movies, puppet shows, and storytellers. The librarian will also help find books on subjects that are of special interest to your child. It's a great resource where kids get to pick out their own books and have access to more than the select few they own.

Preschoolers are ready for more of a story line than toddlers. They love books with repetitive refrains they can chime in on, and they love rhythmic language in poetry and prose alike. Stories about children like themselves or animal stories that are really about children in fur coats provide little adventures that are satisfying and entertaining. Older preschoolers also like books about real people, places, and things. For every age, story books offer new scenarios, themes, and language to play on. And by leading to play, books establish a happy, active connection to the printed word.

Here is a list of books just right for this age group:

- *Freight Train* by Donald Crews
- *The Little Engine That Could* by Watty Piper
- *Curious George* by H. A. Rey
- *Nutshell Library* by Maurice Sendak
- *The Tale of Peter Rabbit* by Beatrix Potter
- *Millions of Cats* by Wanda Gag
- *William's Doll* by Charlotte Zolotow
- *What's the Matter Sylvia, Can't You Ride?* by Karen B. Anderson

- *101 Things to Do with a Baby* by Jan Ormerod
- *Corduroy* by Don Freeman
- *Snowy Day* by Ezra Jack Keats

Instead of the TV-toy connection, parents can help children make the book-toy connection. After sharing *The Little Engine That Could,* why not set up the wooden train set? Or as a follow-up to any story, crayons and paper or puppets can provide a way for the child to relive the story with his own flourishes. Also nice are stuffed animals such as Curious George, Peter Rabbit, and others, which can be found in bookstores and museum catalogues.

Although taped stories should not replace reading aloud, they do fill in the gaps when parents are unavailable and give preschoolers a chance to hear familiar stories as often as they like. You don't have to buy premade tapes. Recording your own readings of favorite books not only is cheaper, but it's a way of having Mommy and Daddy there even when they're at the office or out for the evening. Of course, commercial tapes often have sound effects and music, and some of the best are made by well-known actors and are punctuated with a signal for turning the page of the storybook with which they're packaged. Still, many of the tapes in toyland are recordings of books that sell solely on the glitz appeal of their licensed characters. The stories and illustrations have little to recommend them except that they are instantly recognizable. So be forewarned and take the time to listen to and look at the books.

Recently, Worlds of Wonder's Teddy Ruxpin sparked a craze for animated plush toys. Like Playskool's Casey, Teddy Ruxpin is actually a high-priced cassette player with an animated face that moves as it tells a story. Perhaps for bedtime tales, Teddy would be a hair cozier than Casey the spaceman, but neither is much of a replacement for a warm lap and the spontaneity of a real live storyteller. Neither Casey nor Teddy can answer questions or pause to look at

*Big Bird Story Magic*
*View-Master/Ideal*

details or change his pace or expression. Nor are their storybooks and cassettes, which must be purchased separately, of much literary merit. In fact, many of the books cannot stand alone and tell a story. For the seventy dollars (or more) you spend on a talking teddy, you could buy a library full of wonderful paperbacks—books by Maurice Sendak, M. W. Brown, Wanda Gag, William Steig, and Beatrix Potter, to name just a few. Not only would that library be read and reread, it would be *worth* rereading.

Fortunately, Teddy Ruxpin and Casey are not the only choices. More appealing is Worlds of Wonder's new Mother Goose, whose repertoire includes well-loved folk tales like "The Little Red Hen," "The Three Bears," "The Ugly Duckling," and others. And Sesame Street fans would probably love Ideal's talking Big Bird, who sings and tells stories. Even more versatile is Axlon's Grandpaw Bear, which can be attached to any tape recorder to tell or sing any tape. At times when they simply can't be there, busy parents might prerecord a favorite bedtime story and turn Grandpaw Bear into a stand-in storyteller. While you're considering plush toys that talk, don't overlook Axlon's AG Bear, who doesn't exactly speak, but listens and then answers with a mumble that only a child's imagination can understand. In many ways, AG is the most child-driven of these electronic wonders.

*AG Bear*
*Axlon*

*Talking Mother Goose*
*Worlds of Wonder*

## *Audio-Visuals for Preschoolers*

Preschoolers are especially attracted to electronic equipment, but in some respects such equipment robs them of active play. TV and video cassettes certainly invite more watching than doing, although they can provide information and some scenarios for dramatic play. We can hope that in the future producers will concentrate more on quality and less on licensed characters. But for now, parents need to shop carefully for videos worth looking at once, let alone many times.

If parents really hope to help children develop a love for stories and storytelling, the books, tapes and videos they select should be of the highest possible quality. Since these are materials that will be played over and over, they are taste-shapers for the impressionable early years. Offering less-than-literate stories to preliterate children may fill their time, but it will do little for their expectations and attitudes toward books and entertainment.

## *Computers and Preschoolers*

If you have a computer in your home, there's no doubt your preschooler is going to want to get her hands on it. Young children love to push buttons and make things happen, and any piece of equipment parents use is bound to have special appeal. But if you haven't got a computer, you don't need to rush out and buy one. Your four-year-old will not lag behind her schoolmates if she doesn't have one now. Indeed, most preschool software is highly structured and abstract, emphasizing right and wrong answers and the most simplistic kinds of lessons.

Once the novelty has passed, few preschoolers will stick very long with a computer. Though it is more interactive than TV, it is hardly the ideal toy or teaching tool for threes, fours, and fives, who learn in more direct and active ways.

For software to be useful at all, parents will need to preview it and to evaluate its potential for play and learning. One mother of a three-year-old complained that her little boy seemed more interested in lifting the special overlay for the keyboard that came with a game than in watching the screen or playing the game. In his own way this preschooler was investigating how things worked. For the preschooler, this is where true education begins—with his own curiosity.

Few adults are willing to let preschoolers loose on expensive and delicate computers, nor should they. Instead, they use the computer as an opportunity for one-on-one play. Don't push preschoolers into lessons and high-tech terminology. Use the appropriate language but use it in the context of doing. You can explain the terms *cursor* and *booting-up discs,* but don't overload them with high-tech talk of bits and bytes. Choose software that builds on simple rules and responds entertainingly. Among the best for preschoolers are:

- *Mask Parade* by Springboard: An electronic cut-out book of masks and jewelry kids can wear for pretend play.
- *Muppet Keyboard:* a simplified keyboard preschoolers can use with ease.
- *Number Farm* by DLM: a game of easy counting and matching skills.
- *Stickybear Shapes* by Xerox: easy shape-matching games.
- *Richard Scarry's Best Electronic Word Book Ever* by CBS: reading readiness activities, including both picture-matching and word-matching for the more advanced player.

Less costly but still interactive are various electronic teaching machines like Casey the Robot or the Talk 'n Play tape player, both from Playskool. These machines are often more attractive to parents than to children over time.

*Fun-To-Learn Letters*
*Playskool*

Though they offer novelty, most of their appeal runs out before the batteries. Again, these electronic teachers are programmed to offer narrow lessons with wrong and right answers. They can't answer a child's question or interpret a repeated mistake. So, they're better at drilling than really teaching.

Playskool's new electronic Fun-To-Learn Letters and Fun-To-Learn Numbers, Shapes, and Colors have features that most electronic toys lack. The letters and numbers are there for the child to pick up, feel, and see. This plus the audio response that comes when each piece is placed in its special slot makes some sense. For five-year-olds, the toy's first two levels of play are fine. Level three, however, goes on to initial sounds and spelling, skills more appropriate for six-year-olds. Of course, preschoolers do like a certain amount of repetition and fiddling with push buttons. If you have a limited budget, keep in mind that most electronic toys have limited value and appeal. If you are looking for unlimited play and learning value, save your money for more unstructured toys.

## *SUMMING UP*

By its very nature, play is educational. Yet the child who has eighteen action figures and all the accessories to go with them still has only one kind of play. The same is true of a child with nothing but puzzles and structured toys for "cerebral" training. Your preschooler doesn't need all the toys in this chapter, but a good mix.

In building a collection of toys for preschoolers, ask yourself these questions:

1. Does my child have a variety of physical, imaginative, and intellectual play options?
2. Does she have props for realistic pretend-play or just TV-inspired imaginings?
3. Will a particular new toy extend the use of an old toy or simply be one more of the same?
4. Does the toy I'm considering invite multiple uses?
5. Is this a toy that does the playing while the child watches?
6. Can the toy be used independently or will it require constant adult supervision?
7. If the toy has mechanical novelty appeal, will it still be interesting when the novelty wears off or when the toy breaks down?

# VI. *Toys for the Early School Years: Six to Seven Years*

It's during the early school years that parents and teachers begin to take a dim view of play. Toys and playtime are what you do after your "work" is done. Indeed, one of the threats in the classroom and at home is: "There will be no playtime until . . ." While the importance of beginning school skills is not to be taken lightly, neither should the continued value of play be overlooked.

## *THE NEED FOR ACTIVE PHYSICAL PLAY*

Developmentally, the schoolage child still learns best from active doing. But more often than not, the teaching toys sold for this age group focus on the drill and practice of math and reading materials—a sugar-coated attempt to disguise work as play. Many such educational toys are of limited appeal and value. In fact, children rarely select such toys for themselves or use them independently. To expect children to spend prolonged time with sit-down tasks that are "good for them" is like trying to nourish the body with a bottle of vitamin pills.

For the young child, beginning reading and math skills are just two items on the broad developmental agenda. Equally important is the social business of finding one's place among one's agemates. While home and family continue to be the secure base children return to for comfort and advice, the world of classroom and playground take on new importance. It is in the company of agemates that children test and measure themselves. Making the top reading group may be important and pleasing to Mom, but making it to the top of the jungle gym may be of equal importance to Jessica or Jim. Indeed, it's at the playground in active physical play that children make their early social connections. They don't need to be "the best," but they do need to be able to keep up, to be one of the gang.

If the action in the schoolyard or backyard centers on rollerskates, then skating will take on an importance that needs supporting. The same may be true of riding a two-wheeler, jumping rope, or playing baseball with a proper mitt and ball. Anyone who doubts the importance of physical competence to children has only to watch the seriousness with which they work at self-imposed tasks. Learning to balance on two wheels or two blades, they discover the meaning of the old adage "If at first you don't succeed, try, try again." Providing scaled-down but solid sports equipment is just the first part of what parents can do. At this stage, children may also need a steady hand on the back of the bike seat, tips on keeping an eye on the ball, and a willing mate to play catch with—along with plenty of encouragement. Remember that a young child's sense of physical competence is directly related to his overall view of himself. Too much competition at this stage may diminish the child's budding sense of being able to participate, and even his willingness.

Unfortunately, many schoolage children have little opportunity for active physical play. They ride to school on a bus, sit at a desk for six hours, and spend their evenings watching TV. Often there are no legitimate channels for their energy. A half hour of gym twice or three times a week is hardly sufficient for healthy schoolage children who need daily time to run, jump, and use their muscles as well as their minds. Physical play not only provides a healthy outlet for discharging energy and releasing tension, it provides children with a sense of refreshment and well-being. Yet recent studies indicate that obesity and lack of fitness among six- to seventeen-year-olds is more prevalent today than it was twenty years ago. Children who spend afterschool hours glued to the tube are not just more passive, they're also snacking and being sold on sugared foods and drinks.

Though parents tend to give the highest value to intellectual achievement, it's interesting to note that most children regarded as "gifted" are talented in many areas, not just in cerebral skills. For their children's health and all-around well-being, parents need to encourage and support more physical activity. Supplying the age-appropriate equipment is the first order of business.

## *BASIC EQUIPMENT FOR PHYSICAL DEVELOPMENT*

### *Bikes*

Since competence is central, select a bike the child can learn to ride with ease and safety. This is not the time for buying a bike to grow into. If the child is small, look for a bike to match. An oversized bike only adds to the burden of learning to balance oneself. This first bike doesn't need gears or handbrakes. Keep it simple and inexpensive. In fact, a secondhand small bike is preferable to an oversized new one. Again, a classic red bike will go the distance for younger siblings rather than a fancy job that's gender-specific or faddish. Children of this age come in all sizes, so it's impossible to say what size bike you should buy. But here are some pointers to help you make a wise choice:

1. For the right fit, take the child with you. Forget about surprises.
2. Choose a bike that is easy to straddle without adult assistance. If *you* have to hold it, then it's too big.
3. Your child should be able to put both feet flat on the ground when straddling the bike.
4. Adjustable seat and handlebars will allow for longer use.
5. For safety, bikes should have reflector stripes on the body and fenders. At this age children won't be riding at night, but dusk can set in early and reflectors are a plus.
6. A casing over the chain will help protect shins and pants legs.
7. Find out if the bike comes set up or if you have to put it together. If it comes unassembled, do you have the tools and skills?
8. If you're buying a preassembled bike, be sure it's been put together securely.

Before your child starts riding, establish clear rules about when, where, and how far she may ride. Discuss the reasons for the rules so that she understands why they must be followed.

## *Other Wheeled Toys*

Scooters and large wagons are also popular in some neighborhoods. Neither replaces the bike, but both offer another

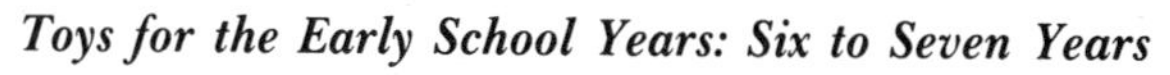

*Radio Flyer Scooter*
*Radio Steel*

way of getting about. So do the big battery-powered cars, tractors, and jeeps. There's no doubt that highly realistic motor-driven vehicles have great appeal to children, as well as to indulgent grandparents and parents. But the fact of the matter is that they have more value as accessories to pretend-play than as aids to physical development. Steer clear of them if you can. Even if money is no object, safety should remain a major consideration: young schoolage children don't always use good judgment and don't need a Porsche that goes ten, fifteen, or thirty miles an hour.

## Skates

Schoolage children enjoy the challenge of balancing on roller skates and ice skates. Indeed, these are often basic equipment for socializing with agemates. Most children will

*Roller Skates*
*Fisher-Price*

need a supporting hand until they get the knack of it. You can assure them that everyone goes through that "pick yourself up, brush yourself off, and start all over again" business.

For beginners, the Fisher-Price roller skates suggested earlier are ideal since they offer a lock that stops the wheels from turning backward. They also adjust to varying shoe sizes and so allow for growth.

If your child will be doing all his roller skating at a rink, you may find rentals or used skates a less costly way of dealing with inevitable changes in shoe size.

Sixes and some sevens may still be more secure on double-runner ice skates. Whether they're learning on a pond or in a rink, you'll find that shoe skates offer more support to wobbly ankles than slip-on adjustable runners. But shoe skates are more expensive and kids grow out of them quickly, so try to rent them or to buy them second-hand.

## *Snow Gear*

For outdoor winter fun, some children are ready for beginner skis. Skis to grow into may be right for your budget but not for your child, since skis that are too long are hard to control and possibly hazardous. A used pair or a rental in the right size may be less glamorous but decidedly more usable. For a less serious but equally playful exploration of skiing, there are relatively inexpensive mini-skis that can be used in the backyard or park.

Sleds and lightweight toboggans are also standard gear for solo and social snow fun. Keep the size scaled down for ease of handling and independence. Bigger is not necessarily better.

## *Balls and Ball Games*

Children of school age start to show an interest in games with rules. That doesn't mean they're ready to play team

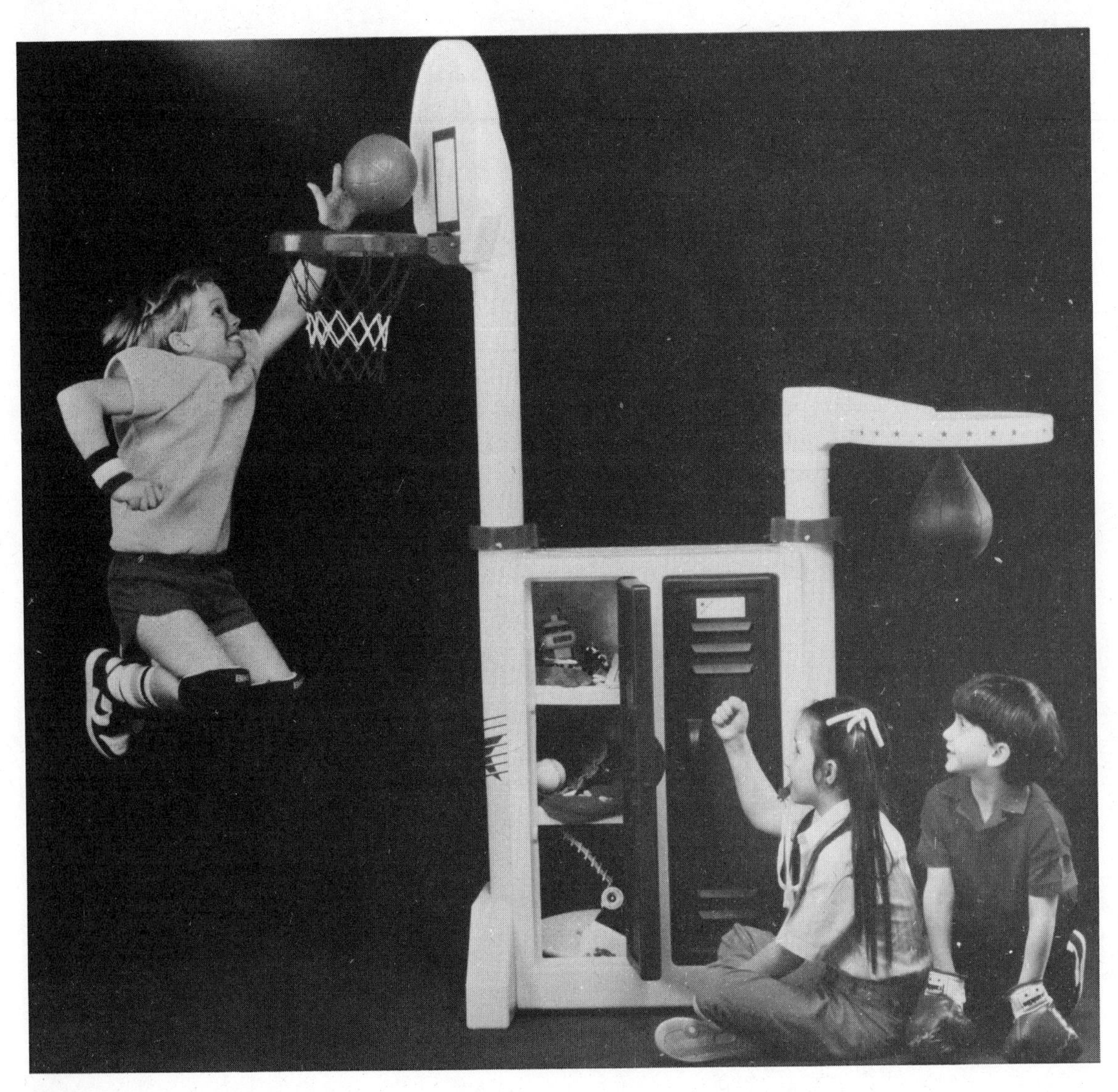

*All Star Sports Center*
*Today's Kids*

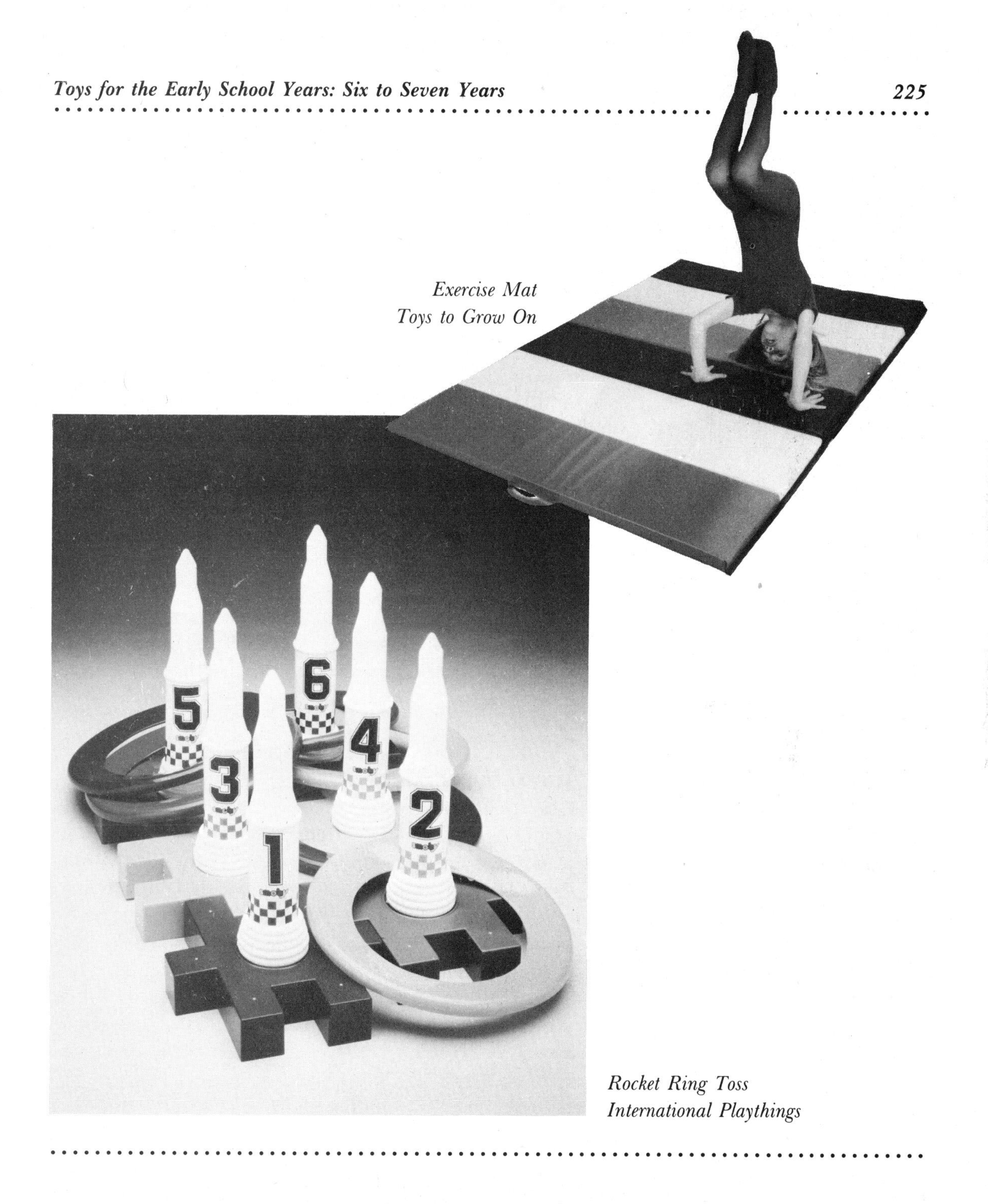


*Exercise Mat*
*Toys to Grow On*

*Rocket Ring Toss*
*International Playthings*

sports seriously or to play all games exclusively by the rules, but they do like playing around with parts of real games and with real-looking sports equipment.

Balls in all sizes and shapes are still favorites for bouncing, tossing against a wall, and catching. Full-size but lightweight versions of soccerballs, basketballs, baseballs, and footballs are easier to handle than the real thing. There's a lot of pretending built into the action. Children of this age aren't ready for much precision hitting, but a game of catch is more fun with the proper costume and accouterments. For safety, avoid standard hardballs and bats; beginners will get the swing of things better with an oversized soft bat and ball. Similarly, they may not be ready for a full-scale game of basketball, but a hoop and ball are a good starting point that can be used solo or with a friend. Today's Kids makes an All Star Basketball Set that can grow with your child. You fill the base with water or sand for stability, and it can be used indoors or out. Also nice from the same maker is the All Star Sports Center with basketball and punching bag attached to a storage locker for stowing gear.

### *Fitness Equipment*

In their eagerness to test the limits of their own bodies and what they can do, schoolage children are especially fond of acrobatics. Many enjoy the structure and sociability of beginner classes. Parents can encourage whole-body action by providing an exercise mat at home and a clear space in which to use it. If space is limited, a chinning bar in a doorway or a climbing rope in the bedroom offer attractive alternatives.

Several toy companies have plugged into the fitness business with weights, hand-press gear, and clothing to wear for exercise. Again, playthings like these have a "costume" or pretend aspect that children may find appealing. But most of it is unneeded. Buy the exercise gear if that's the motivation they need, but whatever activity they pursue, encourage them to invent their own "routines."

## *Other Outdoor Equipment*

For indoor or outdoor action, a ring-toss or target game makes a good choice. Avoid darts with points or suction cups that can be hazardous.

Some older children enjoy the give-and-take of Frisbees and paddle-toss games when they're equally paired. Miniature bowling games can also be used indoors and out. Although scoring is not required, many of these games can be scored if the children involved would enjoy using their beginning adding skills.

Don't overlook the appeal of a jump rope. For this age group few toys hold more physical challenge and social appeal. Once the coordination is mastered, there's a whole language of chants and tricks to expand the repertoire of jumping. In gym class or the backyard, being able to jump rope is an inexpensive way to play and let off steam.

Outdoor climbing gyms continue to hold interest for the sheer physical delight of climbing, together with the opportunities they offer for dramatic play. This is the age when a treehouse at the top of the climbing bars becomes the "clubhouse" or "secret hideaway."

In the same spirit, a playhouse or tent at ground level is inviting for multipurpose outdoor play.

## *Equipment for Outdoor Explorations*

Both sand and water continue to hold an interest for school-age children. A day at the beach is not merely an occasion for jumping waves or splashing. If they haven't already, many children are eager to learn the dog paddle, the dead-man's float, and other rudiments of swimming. Diving underwater for "treasure" is a way of testing and gaining breath control. In the context of diving for pebbles or pennies, children simultaneously develop confidence and the skill needed to become full-fledged swimmers. A diving mask *without a scuba mouthpiece* can encourage underwater looking. Though inflatable floating toys have great appeal, they can be dangerous crutches if children and parents come to depend on them. Lulled by a false sense of security, children may venture into water over their heads in a tube that may deflate or on a float that may slip out from under. Neither child nor parent should rely on inflatable water toys. And remember that swimmers and nonswimmers alike need supervision.

Of course, toy boats, pails, shovels, sandmills, and sand molds all add to the pleasure of designing a splendid castle with elaborate turrets and mandatory moat.

At the beach or in the backyard, schoolage children can be encouraged to notice likeness and difference in the shape, texture, and color of rocks, leaves, shells, and other raw materials of nature. This new attention to details is fascinating and instructive, as this is the age when children begin to enjoy collecting not just store-bought items but also "found" materials. It's fun to sort, count, name, compare, and cart stones, nuts, shells, leaves and pine cones. These collections can be painted, used for pretend, or displayed.

Also interesting for outdoor exploration (and a touch of pretend) are instruments for looking at familiar things in new ways. Tasco makes a Little Looker Pocket Microscope and a domed magnifying glass that are great for examining bugs and other small wonders. A pair of binoculars lends itself to very different close-ups. They're equally good for bird-watching or sighting invading aliens.

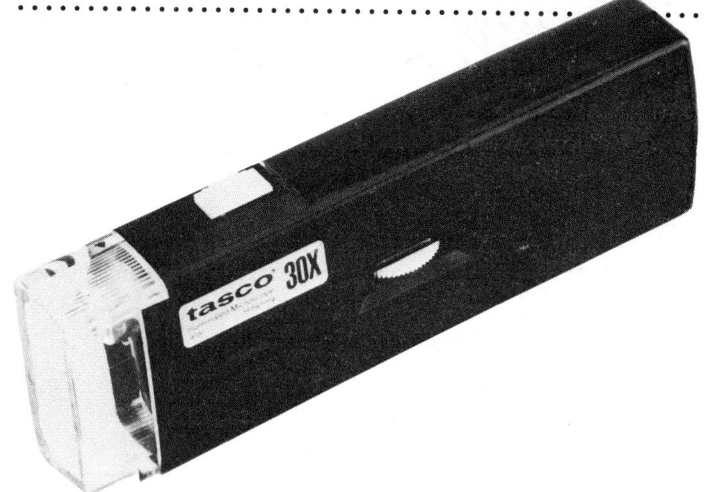


*Little Looker Pocket Microscope*
*Tasco*

Whether you have a backyard or only a windowsill, schoolage children are fascinated with plants. With some well-selected seeds, bulbs, or seedlings, a corner of the family garden or a flower box can produce a showy payoff. Stick to plants that are quick to germinate or sprout dramatically. Lima beans are classroom classics, but string beans, zucchini, and sunflower seeds produce speedy results, too. Paper-white narcissus bulbs don't even require soil. They can be "planted" in water and pebbles and will give quite a show in several weeks. Of course, you can find convenient planting kits at the toystore as well as ant farms and rock collections. But the basic materials in such kits are generally available in your home or backyard for less money.

Bird-watching is another path to active looking that young children enjoy. They're not going to sit for hours with a pair of binoculars, but a feeder outside the window will bring a changing show and small dramas to watch. Children can take some responsibility for filling the feeder; if they do, they'll begin to notice that birds, like people, have different eating preferences and that some are pushier than others. With a simple bird guide, children also enjoy

*Binoculars*
*Tasco*

learning the names of the birds they've spotted, and some may want to start a "collection"—a list of birds they've seen that may grow into a life-long interest.

Helping children connect with the natural world needn't be heavy with information and instruction. In fact, the more active it is, the more memorable it will be. Involvement with birds, shells, stones, or whatever, gives children the kind of hands-on experience they need to become better observers, and it encourages them to investigate and appreciate their surroundings. This is a good time for a small library of field guides that children will enjoy browsing through on their own. The object is not so much to teach names and facts as it is to establish a new way of looking at the world.

## *PETS*

At this stage almost every child longs for a pet. But a pet is not a toy, and young children are not always ready to take total responsibility for the care of a living creature. They're just as likely to overfeed the fish as they are to forget water for the dog. Parents need to be prepared to supervise, and, in most cases, to take on the real responsibility for an animal's care.

Of course, for children a pet is a companion, a friend who won't talk back and who agreeably plays games of toss and fetch, a loving creature who greets them with a wagging tail, happy chirps, or a leap into the lap. The pet's needs satisfy the child's need to be needed, so it's a full circle of caring that adds another dimension to life. For some families, the job of caring for a pet may be more of a hassle than an added pleasure. If this is how you feel, then it's better to be honest and postpone pet ownership until children are old enough to take on the full responsibility.

## PROVIDING TOYS AND TIME FOR SOLO PLAY

Although schoolage children love to be with other kids, there are times when they need a break for woolgathering and playing their own games. Dolls, stuffed animals, art materials, construction toys, books, and puzzles all do the trick. Equipment such as bikes and jump ropes are best mastered in long solo practice sessions. After a long schoolday in the company of others, solo play can provide a welcome change of pace.

Learning to enjoy their own company is no small feat for children. In contrast to their very social life in the classroom and backyard, solo play provides time to pull their own thoughts and feelings together.

## PRETEND-PLAY

Whether they're alone or with others, pretending is still an active form of play for schoolage children.

Some prefer playing games of make-believe using action figures, vehicles, and blocks to create small dramas. Others prefer stepping into roles with their whole being, dressing up and becoming a superhero or a ballerina or the teacher. Most children enjoy both modes of pretend and can enter the world of make-believe through either door.

By six and seven, games of pretend are far more complex, with elaborate scenarios that go well beyond playing house or cowboys. These are the years when cops and robbers, armies, astronauts, and royal princes and princesses hold sway. Dressing up is part of the fun, but a great deal of the pretending is cerebral. Children of this age like realistic props, such as a toy register for a storekeeper or a roll of tickets for a puppet show. Playing with registers and coins is a good way to understand the difference between a dime and a nickel, and how to count change. Paraphernalia For Pretending, made by Creativity For Kids, Inc., contains a wonderful collection of props for pretending to play office, restaurant, store, theater, and more.

Although props are still welcome, keep in mind that sixes and sevens are better able to make one thing stand for another. A stick can become a magic wand, a sword, or anything else, depending on the need of the player. Indeed, if the ability to make such transformations is encouraged, it reaches new levels.

Unfortunately, this is the age when the toymakers' hard sell has enormous impact on pretend-play. Many of the

*Paraphernalia for Pretending*
*Creativity for Kids*

premade characters from toyland and TV are drawn so literally that it's impossible for kids to override the toy's identity.

## *The Social Power of Toys*

For sixes and sevens, the current TV-toys may have some positive social features to balance out the negatives. Although many are violent and ugly, it may be that, in the words of the Transformers' ad, "there's more than meets the eye!"

Elizabeth Gilkeson, former director and senior adviser of the federally funded Follow Through Programs, who has worked with children and teachers for more than fifty years, brings some perspective to the issue: "TV and TV-toys provide a common ground. Years ago, when children lived in certain communities, they might all know about the farm and the work involved. They all had a common background, their homes and families were more alike than unalike. Today they all have a common background of what's on TV."

For better or worse, TV gives children a common cast of characters and situations. "Some of it may seem inappropriate," Gilkeson says. "It may be farther away from reality than it ought to be . . . but Gobots and Transformers are things that relate them to the world of space . . . which is going to be the world of their future."

But there's certainly a price to pay. Comparing the play of earlier generations to the play of children today, Gilkeson asserts that children today can't sustain play for as long as they could when they thought up ideas of their own, ideas that were more within their control and comprehension. "They think they know a lot more than they do about planets and stars," she says.

Gilkeson also believes that "some children can't think up anything as imaginative as they could years ago. They play at getting on and off space ships and have battles in the air. It's as if they read the same book. . . . But in fact they've just seen the same show."

## *Contradictions*

Although many critics agree with Gilkeson, charging that TV-inspired toys actually stifle children's play, reports from parents indicate again and again that children do still make up their own stories. In our survey, the mother of an eight-year-old said that her son frequently created fantasy scenarios that went on for long periods of time. Another parent reported that her seven-year-old uses pop toys for fantasy play with elaborate story lines, a host of well-developed characters with different voices, and imaginative sound effects—"original theme music."

Clearly, there seems to be a disparity between research and the experiences that parents report. It may be that to some extent both parents and researchers see what they want to see. It may also be true that the limited research hasn't managed to gauge the multiple meanings toys play in the lives of children.

From the child's point of view, owning the "in" toy of the moment may have more to do with social needs than with the toy's inherent qualities. If everyone in the second grade is playing with Transformers, the child who has none is likely to feel left out. Even children who watch little or no TV know the cast of the current shows and their toy spin-offs. Indeed, knowing some of the story line and characters may be as basic to the playlife of the child of today as knowing Hopalong Cassidy or the Lone Ranger was to children of earlier days.

The child's eagerness to have what "everybody in school" has may be the first frontier of peer pressure—the place where parental views are in conflict with the new arbiters of taste in the child's life.

Buying a few pieces of the current "in" toy doesn't have to mean surrendering and buying them all. In fact, too much buying can get in the way of the child's ability to create his own imaginative accessories. Elaborate space stations and aircraft carriers are not just costly, they are harder to transform into pretend-play than a child-made ship of scrap lumber, blocks, or cartons. Rather than buying all the pieces, this is the time to supply assistance and encouragement for making a space station of their own, with their own imagination and their own hands.

For most children, the interest in TV-related toys is generally short-lived. This is especially true when children are given other kinds of stories and real-life experiences to play on. It becomes increasingly true when children discover other and more interesting ways to play together. The great challenge for parents is to help children find a balance and move on to more advanced levels of play. In providing a variety of experiences and materials, parents can diminish the dominance of TV-toys' role in their children's lives.

## *Dolls and Plush Animals*

Both boys and girls continue to enjoy soft plush dolls at bedtime or for overnights at Grandma's. Like a security blanket or thumb, these huggables continue to supply comfort as children bridge the worlds of home and away from home.

Small plush toys with brushable or stylable hair are often carried tucked into schoolbags or backpacks for the same sort of security.

Schoolage children also become more involved collecting. Ready-made collectibles like My Little Pony and action adventure figures are used for dramatic play, both solo and social.

By the early school years, the line between boys' toys and girls' toys are more sharply drawn, and children are more likely to play with friends of the same sex. It's as if becoming a member of a same-sex group helps children reinforce their own gender identity. While parents may wish to loosen the confines of male and female roles, children still need to feel they're "one of the boys" or "one of the girls." As a result, children's classic preferences for gender-related toys have changed only slightly. Nowhere is this more evident than in their choice of dolls and plush animals.

Although Tommy may play with his sister's toy dishes and dollhouse, if another seven-year-old boy comes to visit they will not play house. In fact, Tommy is likely to be teased mercilessly if he suggests it. On the other hand, Tommy's sister is likely to have both her "girl" toys and action figures. Not only will she have She-Ra, He-Man's sister, but she may also have Transformers, Gobots, or even GI Joe. In other words, the play possibilities for girls, like the work possibilities for their mothers, have expanded. Indeed, toymakers have plugged into the changing climate and repackaged toy microscopes and telescopes with photos of both boys and girls. However, in the schoolyard or backyard the line between the sexes is as sharp as ever; or, as most six-year-olds will say of the opposite sex: "Girls?" (or "Boys?") . . . *"Yuck!"*

Aside from their plush animals, boys' doll play tends to be limited to action figures. Indeed, the GI Joe Battleship is about the most elaborate boys' "dollhouse" ever created. Boys who played with Cabbage Patch Kids or baby dolls will keep them, but with a low profile.

Girls, on the other hand, get into elaborate doll play at this time. Their dolls are surrogate children who need to be dressed, fed, bathed, wheeled around, and taught. This is the time when dolls with beautiful hair and clothes are enjoyed to the fullest, especially since growing hands can now manipulate simple buttons, zippers, and Velcro fasteners. Corolle and others make oversized baby dolls that can be dressed in real infants' baby clothes. Bottles, rattles, diapers, and carrier are basic gear. Cribs, feeding chairs, and strollers are also appreciated. Sixes and sevens also love girl dolls with jogging, skating, and party clothes that enhance the possibilities for dramatic play.

This is the stage when novelty dolls that creep, walk, talk, wet, cry, and even change hair color are often high on children's wish lists. This year there are Cabbage Patch Kids that talk to each other. Of course you need two for that. But if your child and her friend have one, I'm sure the novelty will be appealing. Let's hope the live kids can get a word in edgewise. I must admit that dolls such as Galoob's Real Baby are amazingly lifelike. Thanks to technology you don't have to pull strings to turn such high-tech dolls on. Just talk to Real Baby and touch her and she moves, talks, cries, and tells "Mom" what to do next. And that's the problem. Like Coppelia, the doll who came to life, this "interactive" doll seems to take over the play. More overactive than interactive, she's the kind of toy that establishes the principle that its the toy—not the child—who does the playing. She's also not likely to last very long. Mechanical dolls generally don't.

*Cherubin*
*Corolle/Brio*

With all this in mind you may still want to buy one. There's nothing wrong with that. An occasional novelty toy is not going to ruin your child's imagination, particularly if the toy will still be usable when its novelty wears thin or it breaks down. But such toys should be the exception rather than the rule.

Although there is some interest in fashion dolls, the younger schoolage child will still have trouble handling the clothes independently. The teenage glamour fantasy fits the older schoolchild better. Again, these are toys better saved for a while.

## *Mini-Worlds*

Boys of this age also enjoy miniature environments for acting out small dramas. Mattel's Hot Wheels Car Wash and Service Station comes in a handy carrying case that makes it easy to tote to a friend's house or store neatly when playtime is over.

Although they're not old enough to handle a racing set with electrical transformers, kids will enjoy Tyco's Speedster vehicles that can race on their Twistrack. The sixteen feet of flexible track can be arranged in any number of ways. Also interesting are simple-to-operate remote-control cars such as Matchbox's Police Car with working lights.

*Remote Control Cars*
*Matchbox*

Both boys and girls like constructing mini-cities with small-scale wooden blocks, vehicles, and animals. Circus lovers will enjoy staging their own show under the big top with the multi-piece circus from TC Timber.

Sixes and sevens are also ready for the type of doll play that centers on the miniature world of the dollhouse. Dollhouses, of course, come in all price ranges, from mansions to basic bungalows. Furnishings can be added gradually for special occasions, rather than buying everything at once. Adding pieces adds interest and invites inventiveness, too. Since a dollhouse is likely to be used over a period of several

*Sylvanian Doll House*
*Tomy/Coleco*

*Puppet Kit*
*Lauri*

years, the business of decorating, rearranging, and home improvement should be part of the ongoing fun.

New and quite charming are Tomy's Sylvanian dollhouses with families of animals and miniature furnishings. The small flocked creatures and their detailed accessories fit right in with the kind of fantasy play that sixes and sevens love. Or, for more classic tastes, see Lundby of Sweden's dollhouse, described in the next chapter.

## Puppets

Another path to dramatic play that schoolchildren continue to enjoy are puppets. At this age they often have an interest in adding not just voice and story, but props and scenery. And schoolage children are capable of making wonderful hand puppets of their own. Lauri's puppet-making kit puts everything they need into a handy package. Of course, old socks, scrap fabric, and papier-mâché for the ambitious add another dimension of play. In making their own puppets, children can tailor a cast to their own story needs rather than adapting their stories to ready-made puppets. Nevertheless, a store-bought cast of human and animal puppets are transformed with relative ease by most children.

Puppets offer opportunities to slip into roles and give voice to feelings that are sometimes hard to express directly. Listening in, parents can often get a better view of a child's thoughts and feelings. Puppets also sharpen language and storytelling ability; and when shared with others, they enhance social skills by underscoring the need to plan and communicate. Often a storybook that's been enjoyed, a TV show, or an actual experience is replayed through puppets. The ability to sequence events and replay a story is no small intellectual task. Indeed, it's directly connected to skills needed for reading comprehension.

*Puppets by Poppets*

Young puppeteers will appreciate a stage, whether it's bought or made. An oversized appliance box works well; and an expandable curtain rod with curtains can turn any doorway into a puppet stage. A premade version from Poppets makes a decorative wall hanging in a child's room and provides storage for puppets as well.

## *Construction Toys*

For some children the process of creating a particular structure is an end in itself. Others build in order to play and use their creations as props, often with action figures.

The child's improved eye-hand coordination allows for more complex and challenging building with larger and more detailed sets of Lego. Legoland theme sets featuring a castle or space station are good choices. Also nice are Tyco's yellow-and-black Construction Vehicle kits that can be used to build earth movers, cranes, and steam shovels. Once built they can be incorporated into dramatic play. And don't overlook a classic set of Lincoln Logs from Playskool.

TYCO
DMT-7
MEN
AT
WORK
BANK
GRAND HOTEL

The wooden logs and slanted roof provide a challenge that requires planning ahead.

Brio-Mec is also interesting. This is a wooden construction toy with nuts and bolts as fasteners. It's something like an Erector Set, but with bigger pieces that young children can easily handle.

For a real departure from the usual plastic bricks, Create It from Wright International, Inc., offers plenty of wheels, tubes, flat plates, and clamps, all of which can be transformed into vehicles or geodesic domes.

Other larger and wonderfully versatile construction toys that sixes and sevens can use indoors and out are Quadro and Playskool's Pipeworks. Kids will need some assistance until they learn how to use the tubular rods and fasteners, which can be fashioned into wheel toys, houses, a puppet stage, a table, and much more. This is an expensive toy, but one with multiple uses over the years. Quadro, the originator, now packs a Mini-Quadro playset in with your big set. The Mini-Quadro can be used to make a model or to construct little houses and vehicles.

OPPOSITE

TOP LEFT
*Create It*
*Wright International*

TOP RIGHT
*Brio-Mec*
*Brio*

MIDDLE
*Construction Vehicles*
*Tyco*

BOTTOM LEFT
*Pipeworks*
*Playskool*

BOTTOM RIGHT
*Lincoln Logs*
*Playskool*

## *Woodworking*

Schoolage children are also ready to handle simple woodworking tools, such as a hammer, nails, a saw, and a plane. With supervision and a supply of scrap lumber they'll enjoy creating their own toy boats, cars, and trains. A bag of preturned wheels and knobs from a craft shop can go a long way in the hands of an imaginative toymaker. Add paint for final flourishes.

In making their own toys, children not only supply their own ideas, they create their own problems to solve. Such constructive play not only builds toys that can be used, but enlarges the child's role in the whole process of play.

A worktable fitted with a vise is a must for sawing. Your hardware store will have a simple saw, hammer, and all the supplies you need. Kids don't need elaborate tool kits. Check the lumber yard for scraps.

## *Art Supplies*

Most of the art supplies listed in the previous chapter will continue to be of interest. Older sixes and sevens will want packs of thinner crayons and are ready to handle them, after a fashion.

Sixes and sevens are ready to experiment with watercolors, which give more translucent effects than tempera paints. A pad of blank paper and tins of paint will occupy seven-year-olds for a long stretch. A small demonstration in how to clean the brush and blend colors will suffice. Start them out with a single pan of six to eight watercolor paints until they've learned how watercolor cakes work. There's no point in buying a twenty-color set and then scolding them for ruining it.

Also interesting are oil pastel crayons that can be blended. Colored pencils and markers of varying thickness are easily enjoyed, too.

Clay and papier-mâché continue to be of interest, as are paste, scissors, and collage materials. Save scraps of fabric, textured paper, and packing materials such as styrofoam blocks and pellets.

Quite apart from getting the creative juices flowing, handling crayons, clay, and paint gives children a pleasurable way of refining eye-hand coordination which carries over into school-related skills. In fact, parents will find that their children frequently experiment with shaping letters and numerals in the context of play. The broad sweeping motion involved in using paints and markers is often more liberating and stimulating for young schoolchildren than the demands of drilling with pencil and paper.

Another favorite medium for art as well as dramatic play is a chalkboard with eraser. Big pieces of colored chalk are less likely to snap or crack, but "school" chalk is a must for the favorite game of playing teacher. One of the great advantages of a chalkboard or an inexpensive lift-to-erase scratchboard is the lack of permanence. "Goofs" can be erased in an instant rather than remaining as reminders of a wobbly *t* or a backward *5*.

During the early school years, children enjoy displaying some of their creations. They take pride in their work and like making gifts for parents and grandparents. A sheet of shelf paper can be turned into a personalized piece of gift wrap for holiday gifts or into a "comic strip" for multiple illustrations that tell a story. Construction paper comes in handy for all sorts of projects, especially cards for birthdays and holidays.

Many of the art-supply kits for this age group help children organize and maintain materials. Holders for crayons and markers keep drawers from becoming clogged and unusable. Keeping things orderly helps to keep them accessible and in good shape. Independence and orderliness are valuable traits for the schoolage child to learn at home.

*Crayola Crayon Case*
*Binney and Smith*

## *Crafts*

With their new interest and pride in creating "end products," schoolage children enjoy trying their hand at crafts. Although their eye-hand coordination is more refined than earlier, they are not ready for projects that demand great patience and painstaking attention to detail. They want results sooner, not later.

Kits with elaborate directions and delicate step-by-step procedures had better wait. For now, small repetitive tasks are more satisfying. A loom for making potholders, a simple jewelry-making kit, or a snap-together model all produce real and usable creations with a minimum need for help.

Sewing is a favorite craft for both boys and girls if the end product is interesting enough. I'm not talking about an old-fashioned embroidery set that requires tiny stitches following a stamped outline. Young schoolchildren will rarely stay with such tasks voluntarily. On the other hand, they enjoy sewing simple hand puppets or "drawing" with yarn using large, blunt embroidery needles and bright heavy yarn. A supply of scrap fabric and yarn, or a puppet kit invite both originality and dramatic play.

Six- and seven-year-olds may also enjoy learning to do needlepoint. Galt's quick point kits come with a small canvas and thick yarn and can be completed quickly. Making something that looks real is no small accomplishment. Steer clear of projects that take weeks to complete.

Paper dolls and activity punch-out books are both entertaining and challenging. Although their play value is usually short-lived, children seem to enjoy the immediate satisfaction of manipulating the pieces. Often the directions and the level of dexterity needed for assembling the pieces will require adult assistance. Some of the handsomest punch-out books would even challenge an adult, so look before you bring home unnecessary frustration.

With their new interest in letters and numbers, children

will enjoy making signs and posters with Childcraft's Rubber Stamp Printing Set. Unlike most printing sets, this one has letters that connect and stamp handles printed so that the child can read what he's setting up before he prints. For a different end product, Galt's Pattern Printing set comes with six intermatching stamps that can create mazes and patterns children can then color or paint.

Also relatively open-ended is the Spirograph from Kenner for creating one's own designs. More literal but still satisfying, Tomy's Fashion Plates offer an introduction to the art of making rubbings. These kits come with multiple costume plates that are raised. It's no substitute for plenty of free drawing, but it does give the child a realistic end product with some decision-making in the process. Like coloring books or color-by-number kits, they're quite satisfying to the child who wants more realism than he or she is capable of crafting. In general, schoolage children do often enjoy the repetition and limited challenge of coloring inside the lines or painting a picture by following a code. It's more puzzle than art. As an occasional form of entertainment, such toys can be satisfying if short-lived.

*Pattern Printing*
*Galt*

*Piky*
*Inova/Childcraft*

## *Puzzles*

Schoolage children are ready for more complex puzzles. For children with limited experience in fitting multiple parts to a whole, puzzles with a frame provide a structure and clues as to where to begin. Many children, however, are eager and able to make the transition to frameless jigsaws with twenty-five to fifty pieces—and ultimately a hundred.

Beginners may need to be given some strategies to cope with so many pieces. Learning to look for the straight outside edges gives them an easy framework. Encourage them to use the picture on the box as a reference point. Without doing it for them, parents can offer helpful suggestions. Unlike older puzzle builders, who generally do a puzzle once and then look for a new one, young children will do, undo, and redo a puzzle many times.

Don't rush to buy the next level of difficulty. Children enjoy and profit from using several puzzles with twenty-five or fifty pieces before pushing on to a hundred pieces. Select puzzles with large pieces that have more surface cues and are easier to manipulate.

Jigsaws are by no means the only kinds of puzzles for this age group. Piky is a delightful puzzle with forty-eight magnetic tiles that can be arranged and rearranged to reproduce picture patterns or create original ones. The colors are vibrant and the challenge is right on target for sixes and sevens. You'll find this in Childcraft. For another kind of open-ended puzzle, consider Kiddicraft's Builda Helta Skelta, a marble-rolling maze that can be constructed in multiple ways.

Multi-sided cube puzzles with pictures also offer a different kind of visual problem. All of the puzzles suggested here not only stimulate visual perception, they also stretch the child's staying power and problem-solving skills. Among the most innovative and handsome puzzle materials are Arcobeleno and Prismatics from Learning Materials Work-

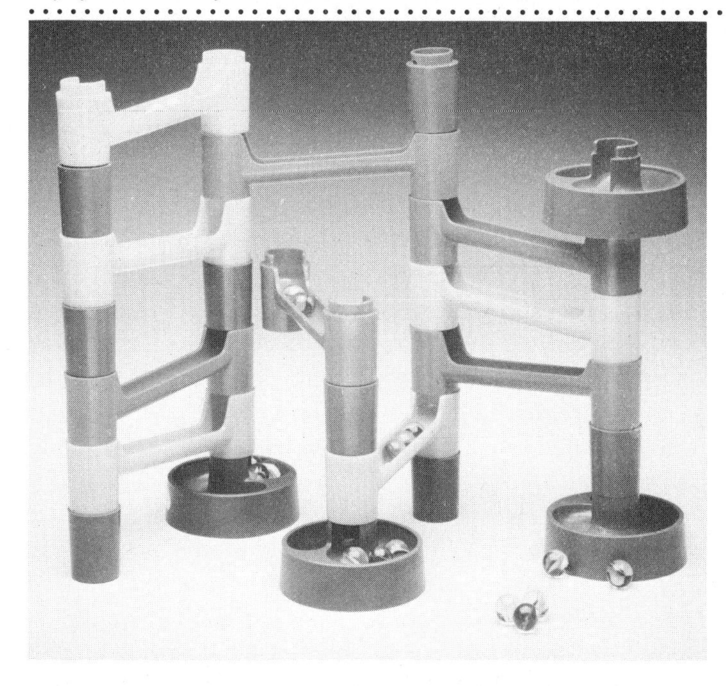

*Builda Helta Skelta*
*Kiddicraft*

*Kids Puzzle*
*Lauri*

*Arcobeleno*
*Learning Materials Workshop*

shop (see Directory). These are beautifully crafted, open-ended puzzles that invite multiple ways of thinking and playing to create designs and patterns.

For children who are grappling with beginning reading and writing, puzzles with letters and numbers are useful. Playing with precut letters and fitting them into frame-boards gives children an opportunity to feel and trace the shapes. This kind of learning offers a sensory dimension that is often more memorable. But puzzles and stick-on or magnetic letters and numerals are not the only path to school-related skills. Lauri's Kids Puzzle calls for looking at details and directionality—the same skills that are needed to distinguish between *b*, *d*, *p*, and *q*. Pictorial blocks and tiles and geometric puzzles also offer playful ways of refining the ability to look at details, fit parts into wholes, find sequences, and reproduce patterns. These skills, closely related to the mechanics of reading and writing, are sometimes more easily learned in the context of play.

## Sit-Down Games

For young schoolage children, the concepts of winning and losing are still difficult. So are games that demand strategy and complex rules. When friends come over, a gameboard is rarely the toy of first or even second choice. Indeed, most such game-playing is usually inspired by an adult, and needs an adult to hold things together. However, a few well-chosen games can offer parent and child a pleasant time together, especially when a change of pace is needed.

Stick to games that are short so that there are several opportunities to win and lose. You don't have to let children win by bending the rules, but there's no harm in giving them some helpful hints when they make an "obvious" mistake. Remember, too, that children of this age believe in the validity of rules for everyone except themselves. They are not above cheating or changing the rules in midstream. Their ethical sense of right and wrong still has a long way to grow. Game-playing is an opportunity to foster its development, but don't expect the lessons to be learned quickly. By seven, children are better able to live with the rules and enjoy the competition—at least some of the time.

For sixes, most of the games in chapter 5 are still appropriate. Stick to games of chance where taking turns and the luck of the draw are the deciding forces.

Many of the pictorial lotto games, picture dominoes, and color and counting games are both fun and appropriate learning tools for sixes. Not only do they expand vocabulary in a playful context, they also build sorting and categorizing skills.

Sixes and sevens enjoy the many variations on Concentration, a game that develops attention to pictorial details and visual memory. They are also ready for simple strategy games such as Chinese checkers, regular checkers, Chutes and Ladders, Bingo, Trouble, and Parcheesi. Toystores, of course, offer an annual flood of new games that are often related to the TV-toys of the moment. With such games, you'll often find that the box itself is more interesting than the game inside.

For sixes and sevens who are well along the road to reading, there are several reading games that are part play, part practice:

- Perquacky (Lakeside)
- Sentences Cubes Game (Selchow and Righter)
- Scrabble for Juniors (Selchow and Righter)

With the first two of these games, ignore the timer and the rules. Just shake up the cubes and let your kids see how many words or sentences they can build.

For another kind of "reading" experience, sixes and sevens can play music by the numbers and eventually move

*Chimalong*
*Woodstock Percussion*

on to real notation with a Chimalong from Woodstock Percussion. This handsome instrument has eight aluminum pipes that produce a true musical sound when struck. The music can be read by colors, numbers, or traditional music notation.

## *Audio-Visual and Electronic Playthings*

At this stage, children will make good use of a sturdy tape machine or phonograph of their own. Their library of recorded music and stories should always be growing. Especially useful for beginning readers are read-along books and tapes. Listening to the story and looking at the illustrations is now enhanced by the ability to follow the text. Indeed, doing so gives young readers good experience at scanning phrases and using the voice expressively.

Unfortunately, with some tape-book combinations, what the narrator reads may not be exactly the same as what is on the printed page. At this stage, the text and the telling should match or the young reader won't be able to follow. Consider, too, the quality of the stories chosen. Since these are storybooks that will be read and heard again and again, the book and the tape should be *worth* repeating. Sadly, this is not always the case. When possible, listen before you buy. Among the most reliable producers of books and tapes are Scholastic, Random House, and Caedmon.

Schoolage children also enjoy recording their own stories and songs, so a tape machine for this age group should do more than just *play* tapes. Several come with microphones that can also be used to amplify the child's voice. These are great fun for dramatic play and "show biz" productions.

Many electronic toys are especially geared to drill and practice school skills. Although most are of limited educational value, they have great novelty appeal for parents, grandparents, and children. From the child's point of view, the fun of pushing buttons and getting instant feedback may be easier on the ego than drills with Mom or Dad. For early math drill, Fisher-Price's Cookie Counter crunches numbers for addition, subtraction, and counting skills. Or look for the Little Professor from Texas Instruments.

Several other electronic toys are less obviously related to school skills, but worth considering. Milton Bradley's Simon is a classic hand-held game that challenges visual and auditory memory. both Texas Instruments and Fisher-Price make junior synthesizers that can be played like a keyboard or programmed to replay original arrangements or songs.

If you own a home computer you'll also want to explore the software designed especially for schoolage children. Although many are essentially electronic workbooks, some are

*Cookie Counter*
*Fisher-Price*

more entertaining and provide the drill and practice that some children need and are willing to do independently. Also consider the more open-ended types of software that invite children to use the computer as part plaything, part learning tool.

Broderbund's *Print Shop* is wonderful for printing up banners, posters, and cards. Learning to use the computer may call for some assistance, but the end products are satisfying. Also interesting for this age group is Springboard's *Mask Parade,* a computer costume box of masks, jewelry, hats and funny feet that the child can design, print out, cut out, color, and use for dramatic play. For projects that require parent-child cooperation, look at Scholastic's *Story Maker,* which allows sixes and sevens to dictate stories that can be printed out with illustrations. With its catalogue of illustrations, this is a great tool for creating lively books about counting, sound, and the alphabet that beginning readers will enjoy.

## *THE REAL THING*

Sixes and sevens have a great appetite for real knowledge, yet they still learn best from concrete experiences. They like the idea of owning a globe, but they can more easily understand a map of their room or of the neighborhood. They're interested in how things grow but they're not ready for wordy explanations. Planting seeds or bulbs gives them meaningful experiences and provides more active learning.

Observing how things change, discovering likenesses and differences, and noticing details are all part of the young child's way of discovering how things work. Shells, rocks, seeds, leaves, and pinecones make great collectibles. For sorting and displaying collections, a field guide and boxes help reinforce the pleasure and learning. Sometimes a premade collection of shells or stones can spark the interest, but found treasures are usually more personally satisfying. Unlike the typical hodgepodge collections of preschoolers, those of sixes and sevens reflect more orderly thinking. Children of six and seven have begun to master reading, counting, and writing skills, and they enjoy using them in sorting and classifying their collectibles. The interest they show in their collections is likely to be more sustained than that of preschoolers.

For young naturalists, unbreakable magnifying bug boxes, tweezers, and bug nets may prove useful. There are also butterfly and ant kits that come with mail-in coupons, although in the summer you'll find everything you need in the way of bugs in the park or backyard. Check museum gift shops or school supply catalogues for these kits.

Magnets, clocks, prisms, flashlights, periscopes, pedometers, and compasses will be of interest to schoolage children. They may enjoy them as objects or incorporate them into dramatic play. Either way, these tools enlarge the child's powers of observation and exploration. Again, you won't find these things in most toystores. School supply catalogues are your best bet.

Cooking under supervision is fascinating at this age. Seeing how butter changes from solid to liquid, how water freezes and ice melts, how cookies change from soft to hard, how water boils and vaporizes—these lessons are easily learned in the kitchen. You don't need to buy any special equipment. Try a simple recipe or encourage your beginning reader to follow the directions on a ready-mix cake box. To you it may seem like work, but to a child there's an element of play with an eatable end product as a reward.

They don't need child-size baking ovens with mini pans and mixes. In fact, used without supervision these can be dangerous. Since you've got to be there when they're cooking, why not use the real thing and give the process more

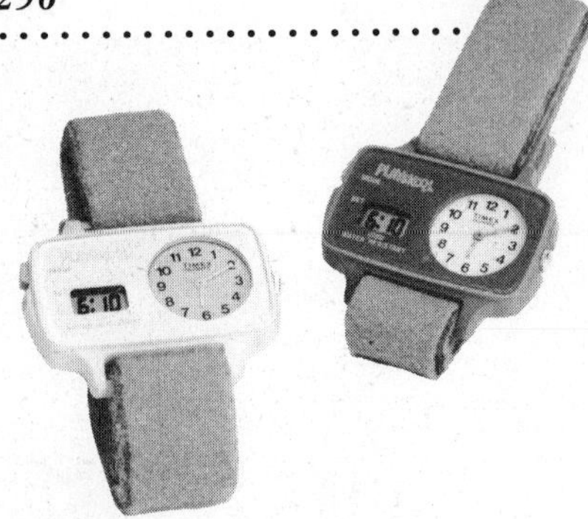

*Timex Teach-Me Watch*
*Playskool*

grown-up status? Preparing a dish for the family meal is a way of giving children a sense of participation in doing something both for themselves and others.

A TV program on wildlife or a trip to your local natural history museum may spark a new interest in playing with models. Miniature animals, especially dinosaurs, have always been fascinating to this age group. Reinforce toy purchases in museum shops with a little background information. Borrow a book from the library and encourage your child to draw or sculpt original creatures from clay.

The distinction between playthings and working equipment is sometimes hard to define. Early schoolage children are no longer content with plastic models that look like the real thing. They want a watch that runs, a camera that takes pictures, a pair of binoculars that really brings things closer. How they use or care for them is another matter. They are not likely to remember a long list of prohibitions, so fully functioning equipment made especially for children is your best bet.

In this department Fisher-Price leads the way with a sturdy camera that's simple to operate and relatively inexpensive. Binoculars and walkie-talkies are also ideal for pretend-play or practical use.

Since sixes and sevens are learning to tell time, they especially enjoy a wristwatch of their own. Again, Fisher-Price has an appropriate waterproof watch that's practical and less cutesy than most. Also interesting is Hasbro's watch with dual digital and analog readouts. Although the digital readout can be something of a crutch before children are comfortable telling time, it serves as a useful cue. And since sixes and sevens rarely get beyond reading the hour, half hour, and quarter hour on a regular watch or clock, the digital readout may serve as an introduction to the next level of precision.

OPPOSITE
*Walkie-Talkie*
*Fisher-Price*

## *SUMMING UP*

In looking at the variety of toys and playthings for children in the early school years, it's easy to see that action figures represent only one of the dimensions of their playlives. No child needs all the toys suggested here. However, when given a variety of materials and experiences to feed their play, sixes and sevens will not be overly influenced by the hard sell on TV. Nor should parents believe that only school-related toys are worthwhile. Indeed, all the playthings in this chapter have broad educational value to the child who has much to learn as a physical, social, and intellectual being. For parents, one of the big jobs is keeping that excitement about learning alive with play materials that are truly engaging.

# VII. *Toys for the Middle Years: Eight to Eleven Years*

These are your child's final years in toyland. None of us outgrows our need to play, but the kinds of playthings we enjoy keep changing. Some of those changes begin to emerge during these middle years of childhood. Physical play takes on a new seriousness, and with it, a need for real equipment. Often such items are better bought in a sporting goods store than at the toystore. In a similar fashion, serious arts and crafts materials may be more easily found in art supply and craft shops. Likewise, model builders will graduate to hobby specialty stores. In fact, this moving up to "real" tools and equipment goes along with the child's growing abilities and expanding interests.

Yet the same child who's ready for a real basketball and hoop may still long for the latest Transformer or Barbie

*Pogo Bal*
*Hasbro*

doll. The middle-years child is often a wonderful, contradictory blend of past, present, and future.

For many, these last flings in toyland are marked by a huge appetite for big ticket items. The fanciest and most expensive doll, game, remote control car or racing car set becomes the most desired toy. It's as if only the biggest toy matches their image of their almost grown-up selves.

During these years, parents have a decidedly smaller role in their children's playlives. Children may enjoy and benefit from an occasional board game, cooking project, or a game of catch with parents, but playing with friends is number one on the agenda.

Indeed, finding acceptance and a comfortable place among one's agemates is one of the central developmental tasks for the middle-years child. While the younger child's self-image is largely shaped by parents, the schoolage child looks to his peer group for a large measure of his self-esteem. In dozens of small but meaningful ways the child tests himself in the classroom and at the playground. Although parents' opinions are still important, there is now a new cast of authorities and opinion-makers who play a large part in the child's everyday experiences.

Since "belonging" is so important, these are the years when clubs and teams—really mutual-admiration societies—are swiftly formed and abandoned, often in the same day. The function of the club is not only to be "in"-cluded but to exclude others. The main activity is frequently talk—or more precisely, arguing. Middle-years kids get caught up in debating who's going to be allowed to do what, not to mention when, where, and why they're allowed to do it. Indeed, children of this age often spend more time talking than doing.

Sports and games with rules become increasingly important. Softball, soccer, touch football, and basketball are played with negotiable rules. Being a member of a team provides another kind of much-desired belonging. Few children are ready to be superstars. In fact, being part of a team offers more insulation and comfort than having to carry off something by themselves.

On the other hand, this is also the time when many youngsters get started with sports that will eventually lead to solo performances, such as skating, gymnastics, and swimming.

In their eagerness to try out new and varied activities, middle-years children tend to flit from one enthusiasm to another. This week's desire for ice skates may be displaced by an urgent need for skis next week. Today she wants piano lessons but next week it's ballet classes. This appetite for new activities often overshadows prolonged interest in any one thing. This stage represents a new kind of exploration, a way of testing their talents, likes, and dislikes. For parents it can be frustrating and costly. When possible, renting equipment or buying (or borrowing) second-hand skis, skates, and musical instruments helps you cut your losses if you child's interest is short-lived.

## *Collecting*

Middle-years kids are great collectors. Many are into collecting stamps, postcards, chewing gum cards, menus, buttons, and soda cans. Comparing collections with friends, swapping and keeping track of what they have and what they want is all part of the fun. Although some collect by-products such as stamps and political buttons, many are heavily interested in "buy products." GI Joe with all his paraphernalia is a great favorite, chiefly with boys. For girls, this is the age of Barbie, and the more accessories the better. This acquisitive mania for ready-made collectibles is usually short-lived. In fact, parents can sometimes enlarge and refocus the child's view of "collections" by suggesting other interesting collectibles such as shells, rocks, coins, and other "found" objects.

On a more interactive level, this is also the age when penpals offer another sort of "collecting" and access to a wider circle of agemates.

It's during these years that children become far more critical about TV ads and toys. Compared to their younger brothers and sisters, they are a tough audience. In answer to our survey question, "What do you think about TV commercials?" they wrote:

> "I think that the people who advertise don't know where to put them. I mean, who watching the 'Transformers' [show] would want a Barbie doll?" (Ten-year-old boy.)
>
> "I hate them, but some of the best toys I have have TV commercials." (Eight-year-old girl.)
>
> "They don't convince me!" (Eight-year-old boy.)
>
> "It lies!" (Eight-year-old girl.)
>
> "Some are O.K.—but most are rip-offs." (Ten-year-old boy.)
>
> "Usually they make toys look better than they are." (Nine-year-old girl.)
>
> "I *hate* them . . . I like to buy toys because I like them, not because the ad made them look good." (Eleven-year-old girl.)
>
> *"Bad* . . . only sometimes *good."* (Eight-year-old girl.)
>
> "I don't trust them. Some commercials are false and they rip you off." (Eight-year-old boy.)

Clearly, living in a consumer-driven world means that at an early age children begin to doubt what they see and hear.

While we might wish that TV advertising was not so relentlessly targeted at children, there is comfort in knowing that by eight and nine, kids become more critical of the media's hard sell.

## *Social Pressure*

The pressure to buy, however, does not come solely from TV. As social beings, children of this age are tuned into the fads of the moment. Even as they are becoming more skeptical about TV-toys and ads, owning the "in" toy continues to have more to do with a strong interest in belonging than it does with a strong interest in the toy itself. From the child's point of view, owning the newest action figure represents the ticket to group action in the lunchroom or at the playground. By recognizing the strength of their children's need to belong, parents may better understand the urgency of some requests.

Think back to your own childhood. For you, it may have been a Hula-Hoop or Frisbee or even the original GI Joe or Barbie. For your parents, it may have been jacks, marbles or a Dick Tracy Secret Decoder Ring. The point is, for each generation, childhood has its own tokens of belonging, which provide a point of departure for joining in and playing together.

For today's children, the robots and their intergalactic wars seem to fill the bill. Traditionally these have always been the years when children played elaborate chase and escape games.

As one mother put it, "We played army, cowboys, and cops and robbers. How's it different? I mean, what's the harm?"

To some extent she's right. It's not so remarkable that children can play war better than peace. Casting is simple. There are the good guys and the bad guys. The hiding, catching, and rough-and-tumble dramas all provide an outlet for tremendous action. By contrast, how do you play peace? Even world leaders don't seem to know how.

Nevertheless, to compare today's TV-related toys with an earlier generation's games of cowboys or army is like comparing Huck Finn's raft to an ocean liner. They're vessels that travel on water, but the similarities end there. Both the quantity and quality of current TV and its spin-off toys are significantly different. The Lone Ranger or Hopalong Cassidy trapped the enemies of justice in the old corral by outwitting them with brains as well as bravery. Today's TV heroes rely either on hi-tech weapons or magical powers. Cast in fantasy, the stories are simplistic and the characters have no dimensions. They are either all good or all evil, with no nuance or shading. Indeed, the good guys use some of the same strong-arm techniques the bad guys use. While many high-tech fantasy figures seem to inhabit a post-nuclear world of the future, Rambo, GI Joe, and Delta Force are more like spin-offs of the evening news—what you might call prenuclear patriots. Although they may reflect the state of life in the late twentieth century, that may not make them appropriate toys for children. But the fact is, they're on the market, and likely as not in your children's hands.

We do know that children who watch more TV not only play more aggressively, but they become more aggressive consumers. In fact, one study shows that when they describe what sort of adults they'd like to become, heavy viewers list the things they want to *own* rather than the kinds of people they want to *become.*

Parents can find some comfort in knowing that:

- Children's fascination with faddish toys is usually short-lived. Like their love of comics, Nancy Drew, Tom Swift, and bubble gum, this too shall pass.
- By this age, children also have a better grasp on what's real and unreal, probable and improbable. For the

child who's sorted out that much, the TV-toyland world of mayhem is probably of little harm.

- Children's interest in mutants and warriors is not a signal of a sick mind in the making.
- Some of the fascination with uglies is their shock appeal. It's precisely because parents don't like them that kids love them.

Parents can certainly do more than wring their hands and surrender. They can provide children with multiple alternatives. Saturday morning need not be given over to marathon viewing and follow-up buying. There are bikes to ride, hikes to take, museums to visit, and concerts, movies, ballet, and theatre to attend. There are arts and crafts classes to join, libraries to visit, and books to read and share. There are even quality TV shows to be watched and enjoyed—if someone takes the time to point them out. In other words, even with the preponderance of poor programming for children, parents can do a great deal to bring balance to the kinds of entertainment that engage children.

The challenge for parents is to provide bridges to multiple forms of entertainment. Since children of this age are able and interested in learning to use tools, why not provide the raw materials for building their own space station or battleships? Rather than buying all the accessories, supply them with fabric, wood, paint, and a helping hand. A puppet show for friends or family, or homemade gifts that are unique can be joint ventures leading children to creative and imaginative play that matches their abilities and enlarges their interests.

## *"But It's My Money!"*

By this age many children have money of their own to spend. Birthday and holiday gift money and allowances often lead to disputes over how such money is to be spent.

Just about every parent has heard the words, "But it's my money!" Within limits, there can be some value in allowing children to buy what you consider a poor choice. You can voice all your reasons and make other suggestions, but in the long run children need some opportunities for making mistakes. They also learn from living with the consequences.

## *Setting Limits*

As big and independent as children like to think themselves, it's still parents who need to define what's safe and unsafe, acceptable and absolutely prohibited.

For example, nine-year-old Larry's closest friends are enthusiastic skateboarders. Larry's mother is a doctor who refuses to buy a skateboard. Why? Working in the emergency room of the hospital she's seen two deaths and far too many serious injuries from skateboards. She knows that Larry takes plenty of teasing from his buddies about this, but she remains firm on the issue. As she said, "He seems to accept it and understands why I've said no."

Larry may not be pleased with his mother's decision. But by confronting the issue of safety rather than criticizing his friends, she has defused the conflict. By taking a firm stance, she is helping him learn to balance the interests of the group with his own best interests. Though children of this age group may argue otherwise, they still need parents who set limits and protect them from their own lack of experience and judgment. Whether it's skateboards, all-terrain vehicles, or BB guns, it's parents who can help children avoid becoming enslaved to a group mind-set.

## *Toys and Gender*

Both boys and girls of this age continue to play games of pretend. In fact, some of their more serious pursuits in sports or theatrical productions have a quality of pretend-play to them. Real athletes, singers, actors, and comedians—along with their mannerisms, equipment, and style—present models to which kids are keenly attuned. At this point, children pretend not only by using miniature dolls, but also by role-playing with their whole being.

Although some girls have little interest in dolls, most are still delighted with dolls of all sizes and types. Elaborate baby dolls, fussy girl dolls, as well as fashion dolls, all with tons of accessories, are greatly desired. Barbie and her paraphernalia satisfy the desire to collect and propel age-typical fantasies about being grown-up.

Several of the parents surveyed indicated some reluctance to buy into the sexual stereotype that Barbie and her glamour represent. Yet parents are sometimes surprised at the quality of play Barbie provides.

> "Barbies were my dilemma because the girls wanted them—and their clothes and and accouterments. I would give in against my better judgment and was always proven wrong in that it led to the best fantasy play."

> "I once thought of Barbies as disgusting, but they create incredible structures and fantasies that go on for hours and from room to room . . . much as paper dolls had been for my sister and me."

Elaborate dollhouses are also used for pretend-play and craft-related activities. Again, the joy of collecting, creating, and dramatizing all come together under one roof. For some children this is the beginning of a long-term hobby in collecting or making miniatures. Lundby of Sweden makes a preassembled, prewired dollhouse with wooden furniture that's built to take active play.

In addition to the hottest spin-offs from TV and the movies, both boys and girls continue to enjoy soft dolls. Some

*Barbie and Ken*
*Mattel*

*Dollhouse*
*Lundby*

*Ophelie*
*Corolle/Brio*

*Ko Ko*
*Dakin*

are used for dramatic play, others as comfortable companions at bedtime.

Although some prefer small and varied critters, others prefer to collect one favorite animal. Often a fascination with horses or dogs determines the collection. However, many children of this age group like zany-looking animals. Maybe such far-out dolls look less babyish and more sophisticated. It was this age group that first adopted those homely Cabbage Patch Kids.

Of course, boys of this age group are apt to keep a low profile when it comes to dolls. In an interview with a ten-year-old boy on vacation in the Caribbean I asked, "What toys did you bring along?" He told me about his action figures and a board game. When his sixteen-year-old sister reminded him of the Pound Puppy he had tucked into his backpack, his face reddened.

Girls, though, have expanded options. In talking to a group of nine-year-olds about their toy preferences I asked: "Is there really such a thing as girls' toys and boys' toys?" Their answer was a resounding "Absolutely!" But here's what they went on to say:

> "A boy might want a Transformer but he wouldn't want a Barbie doll."
>
> "A boy might play with GI Joe but he wouldn't want to play with Strawberry Shortcake."
>
> "But a girl might play with Transformers or GI Joe. Some girls do, and they also play with Barbie."
>
> "I don't think boys would like Barbie, but they could play with My Little Pony if they want to. I mean, there's no rule about it."
>
> "On the My Little Pony commercial you usually see girls, and you see boys playing with GI Joe. But I think toys can be sort of like co-ed—anyone can play with any toys they like. People are all different and they can have different opinions."

As we've already seen, these expanded possibilities echo the new roles women and men are playing in the work force and at home. Today's mom may be a fire fighter, a lawyer, or a police detective, but in taking on roles that were formerly man's work, she has not given up her nurturing role as mother nor her gender-related appearance. Today's dad may take a more active role in sharing responsibility in childcare, cooking, and household chores, but neither has he changed his look. The point is, we have not all ended up in the unisex jumpsuits that designers envisioned in the sixties. And children's toys reflect as much.

## *Show Biz*

This is the stage when the show-biz bug really hits. Some children find an outlet in performing magic tricks. Others sing, dance, or direct and star in variety shows. Still others set up carnivals or "money-making" flea markets. Quite often the middle-years child dreams bigger dreams than he can carry off without a bit of parental aid. For parents this is an opportunity to lend a hand. It may mean a trip to the lumberyard or the donation of an old pair of drapes for curtains.

For junior theatricals, a tape recorder or phonograph are useful. Some come with microphones that can be used for singing along or alone. Often children begin lessons on real musical instruments at this age. If there's no room in the schedule or budget right now, the Chimalong and some of the small electronic instruments mentioned in the previous chapter may satisfy their interest in composing or playing familiar tunes.

If there's a computer in the family, the *Bank Street Music-Writer* offers another route. While the tone quality leaves much to be desired, it is an acceptable alternative to more costly instruments.

For theatrical performances and magic tricks, children like costumes and makeup, The local library will have magic books or you will find a book and materials for tricks in Battat's Hocus Pocus Set. For aspiring jugglers, Chasley makes glow-in-the-dark Shooting Stars, or a set of veggies called Tossed Salad, lightweight objects to juggle packaged with step-by-step instructions. Adult clothes and scrap fab-

*Tossed Salad Juggling Set*
*Chasley*

*Video Camera*
*Fisher-Price*

ric are often more attractive than ready-made costumes. As for makeup, I prefer providing real nonallergenic makeup to the toystore variety.

Fisher-Price has introduced a video camera designed for kids. Compared to the cost of the real thing it's a bargain, but don't expect color or a very sharp picture. This is a toy, but one that kids can use for taping friends and family.

## *Puppets*

As always, puppets provide another route to pretend-play. By now children are quite able to create their own puppets. With papier-mâché and fabrics these puppets are often works of beauty, wit, and imagination.

For some, learning to manipulate stringed marionettes is a challenge, as is the business of playing out familiar and original stories.

## *Transportation Toys*

Although many children continue to collect miniature cars and trucks, this is the age when children especially enjoy remote-control vehicles, electric trains, and racing cars. Very often the interest in such toys is on-again-off-again. They will be used intensely and then ignored for weeks or months. However, for some kids this is the beginning of a lifelong fascination with motorized miniatures. Rather than making a huge investment, start out with a relatively basic set. You can always add pieces if the interest is sustained. You can also encourage and help them make some of their own props.

Lionel's Nickel Plate Road Set features a die-cast engine that pulls five units. It's a classic that looks a lot like the one Dad remembers fondly. He may not be so fond of the price. For a more moderately priced set, consider Lionel's Can-

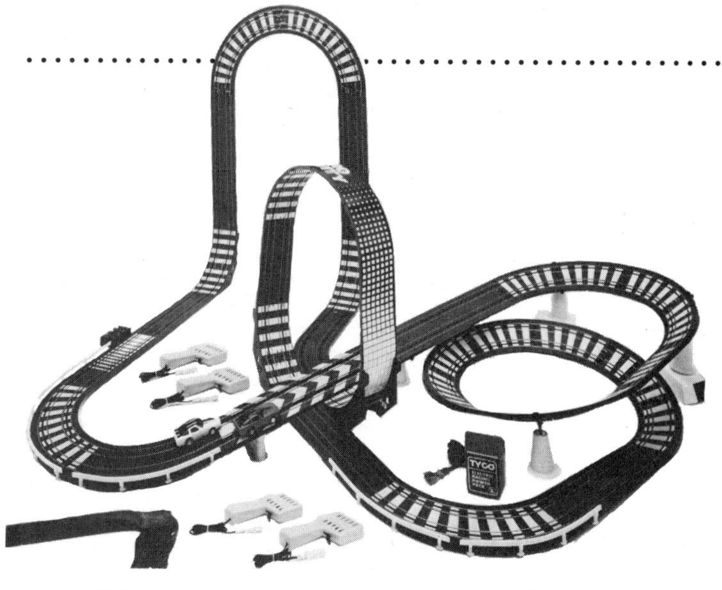


*Zero Gravity Cliff Hangers*
*Tyco*

nonball Express, with a steam locomotive, five cars, and a bridge.

For racing car enthusiasts Tyco's Zero Gravity Cliff Hangers has cars climbing walls, racing through loops and spirals, and even speeding upside-down.

To use such toys often requires the space to leave the tracks and props out for a while after having put so much thought into setting them up. By defining an area for such use or providing a surface that won't be disturbed you'll avoid a lot of conflict and won't end up tripping over pieces.

*Remote Control Racer*
*Skilcraft*

Battery-operated remote-control vehicles are another favorite with this age group. They come in a wide range of styles and prices. Set a budget and go shopping together. Kids generally have strong style preferences and want a say in the selection.

## Sports Equipment

As kids become more involved with sports, they want regulation-size equipment. Depending on the season and the neighborhood, middle-years sports most often include:

- baseball
- basketball
- football
- soccer
- ice skating
- skiing
- roller skating
- sledding
- tennis
- badminton
- table tennis
- swimming
- kiting
- croquet

Few pieces of equipment are more central than the two-wheeled bike. By eleven or twelve, most kids can handle a full-size adult bike with gears. Smaller and younger riders may need a transition bike, since a bike that's too big remains a potential hazard to safety. Kids, of course, want what's fashionable and what their friends are riding. But a bike that doesn't fit is an invitation to the emergency room. Go to a bike store rather than a toy store. They'll have more choices and help you better to find the right fit. And be sure to take your child along rather than surprising him. Also important at this stage are firm rules about where and how the bike will be used. Reflectors on bikes should be used even if the bike is not to be ridden as a rule after dark.

Children of eight to twelve are *not* ready for motorized bikes or all-terrain vehicles. Their proclivity for taking dares along with their lack of judgment make engine-powered vehicles a dangerous choice.

In the toystore you will find some sporting equipment for the middle-years child. The Nerf Ping-Pong Set from Parker Brothers can be used on any tabletop. So you don't need a rec room to have a game. Ohio Art's Badminton Set also comes with rackets and balls for miniature tennis matches. It's not the same as having your own tennis court, but if you've got a backyard or a driveway, it's not bad. The bases for the net can be filled with water or stones for stability. Or for some vigorous tests of their own stability, consider Hasbro's Pogo Bal for jumping or an exercise mat for tumbling.

*Kites*
*Go Fly a Kite*

*Radar Ball*
*Skilcraft*

*Volleyball Plus*
*Ohio Art*

## *Crafts*

Given their new and more refined dexterity and their love of creating end products, crafts projects hold a special appeal for this age group. Sometimes their interest will grow out of a fascination with the past. As school children begin to study history and the way people lived, they enjoy the opportunity to try their hands at weaving, candlemaking or woodworking. It's a way of making history come alive in active and meaningful ways. A visit to a craft fair, an art museum, or a historic restoration may inspire both children and parents in cooperative ventures.

Since children have little patience for drawn-out projects that take weeks or months to complete, start with simple projects. Before investing heavily, remember that children of this age are still far more interested in exploring than in giving painstaking care to perfecting their techniques.

This is the time for kits that come with all the tools and materials needed for a single task. A small needlepoint kit, a leather tooling project, or a simple loom for weaving will

RIGHT
*Needlepoint Kit*
*Galt*

*Weaving Loom*
*Fisher-Price*

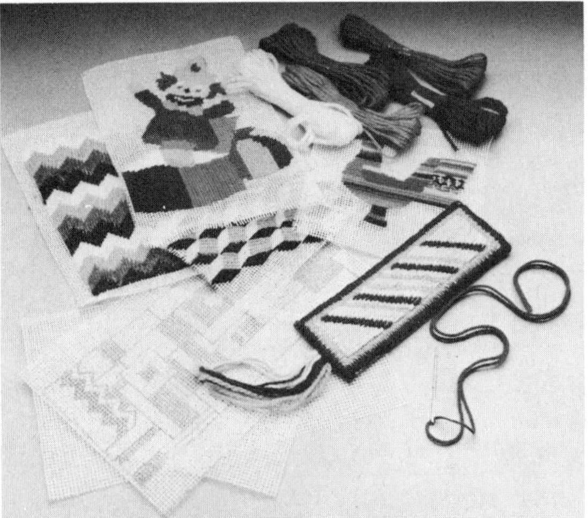

offer a way of trying on an experience without too much of an investment.

Remember, too, that with some kits you'll need to be available as assistant or adviser. Safety rules need to be established for using tools and storing materials out of the reach of younger members of the family. Naturally, if the craft calls for hot wax, ovens, or irons, adult supervision is a must.

Working at a task that requires step-by-step execution, concentration, patience, and dexterity adds up to a rewarding experience. Here are some of the beginner craft kits that middle-years children will enjoy:

- weaving looms (Fisher-Price)
- woodworking kits (Woodkrafter Kits, Inc.)
- jewelry bead kits (Schowanek of America)
- needlepoint (Galt)
- bead crafts (Hama)

Don't overlook the possibilities of craft projects with found materials:

- carving soap
- tie-dyeing
- batiking
- sewing doll clothes
- making puppets
- making masks

## *Art Materials*

Although all the art materials suggested in earlier chapters continue to be of interest, children of this age are ready for some new materials. For drawing they'll enjoy trying pressed charcoal, oil pastel crayons, colored pencils, pen and ink, and calligraphy tools. For painting they're ready for large sets of watercolors and tubes of acrylic and oil paints. Binney and Smith's Creative Lettering Set gives them simplified markers to try their hand at calligraphy.

*Creative Lettering Set*
*Binney and Smith*

And the same company's Designer Kit for Vehicles offers doable experiments with drafting tools.

Very often children of this age group become critical and self-conscious about their own creations. Though they love to draw elaborately detailed pictures, their skill in reproducing reality usually falls short of their expectations. Parents can help children by encouraging them to look at how light and distance affect the size and color of things. They can also point out how artists take liberties with what they see. Museums, art books with reproductions, and even commercial art in advertisements are good points of departure. Children should also be encouraged to continue exploring more plastic art forms such as clay, collage, and papier-mâché which lend themselves to freer expression and invention.

Remember also that children at this age often have art experiences in school that they like to follow up at home. While the school experience is usually limited to thirty or forty minutes and tends to be teacher-directed, art work at home can be self-directed. It gives kids a chance to hone skills and also find a nonverbal but meaningful form of self-expression.

## *Construction Toys*

Middle-years kids usually graduate to more complex model building. Traditionally this is the time for Erector Sets, with their endless possibilities for multiple building tasks. Advanced plastic building sets like Lego and Fisher-Price's Construx, offer similar challenges and satisfying results. Both of these have motorized accessory kits that kids can use to add a new dimension to their constructions. Adding new functions to an existing set extends the life of the toy and its usefulness. Play-Jour's Capsela, with its futuristic look, provides both models and open-ended building possibilities that introduce kids to concepts in physics. With Fantastix from Integrity Design, nine-year-olds (and up) can build amazingly strong geometric structures from rigid struts and stretchy tendons. You'll find Capsela, Fantastix, and other science-related toys in museum shops or in the Discovery Corner Catalogue in the Directory.

For more structured building experiences, children also enjoy the realistic details of models. Start with less complex models with big pieces that snap together and demand less precision. As children move on to more detailed models they may need assistance in dealing with step-by-step directions, small parts and techniques in handling glue, decals, and paint. Remember, too, that these materials must be used safely. The room must be ventilated, surfaces need to be protected, and works in progress need to be out of the reach of younger siblings.

The end products of constructive play often fit right into the middle-years child's interest in collecting. Their efforts are often a source of great pride.

Older children in this age group will also be fascinated by Spacewarp from Ban Dai. Putting together this roller-coaster is a challenge that required plenty of patience, but after using the suggested layouts, kids will enjoy designing their own roadways.

*Capsela*
*Play-Jour*

*Construx*
*Fisher-Price*

*Fantastix*
*Integrity Design*

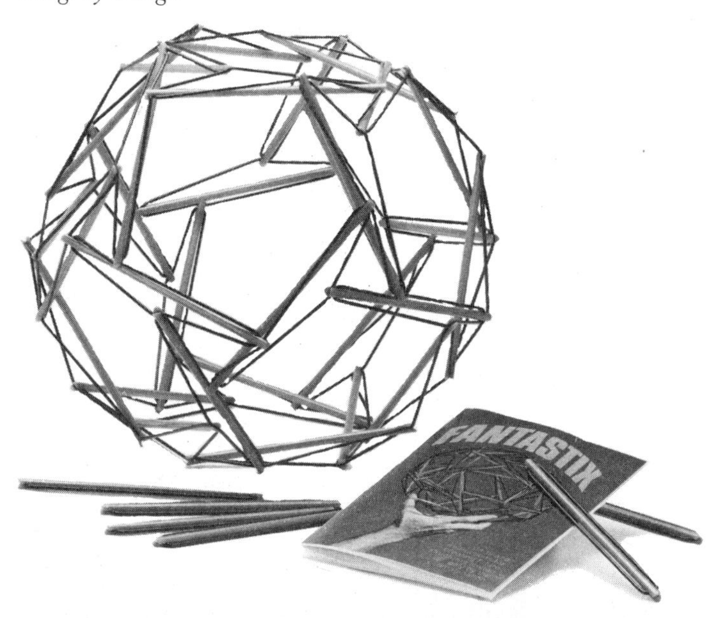

*Dinosaur Model*
*Monogram*

*Spacewarp*
*Ban Dai*

## *Cooking*

Although for you, cooking may be more of a chore than a pleasure, for boys and girls it's a multifaceted experience. Most of the cooking ovens designed for crafts or for children's cooking are essentially a waste of money. In fact, some may be downright dangerous. If you've got a working stove you don't need a toy stove that may give kids a shock or tempt them to experiment when you're not available. Establish the rules of the kitchen and teach them to use real tools with supervision.

Following the directions on a ready-mix package of soup, pudding, cake, or rice gives kids a combined reading and science experience. Or use a recipe book and measuring tools to whip up some edible treats. In either case, kids are learning about putting things together in a sequential order and discovering that numbers and fractions have a practical purpose in the world outside of school. In creating something that the whole family can enjoy, the middle-years child can take pride in doing something that's both fun and useful.

## *Science Materials*

As boys and girls begin to grapple with more abstract ideas, they continue to benefit from materials that are more concrete. Many science kits give children a chance to discover how things work. They bring big ideas into child-size explorations. Words like energy, electricity, and air pressure take on more meaning when they are translated into seeable and doable experiments.

Some of the most interesting are:

- potato clock (Skillcraft)
- telescope (Tasco or Bushnell)
- microscope (Tasco or Bushnell)

*Telescope*
*Tasco*

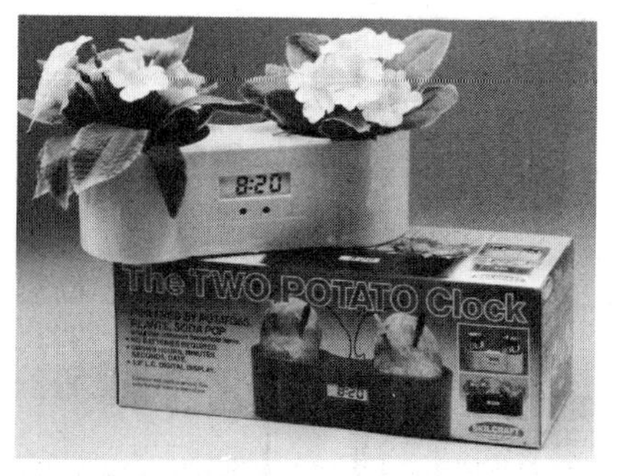


*Two Potato Clock*
*Skilcraft*

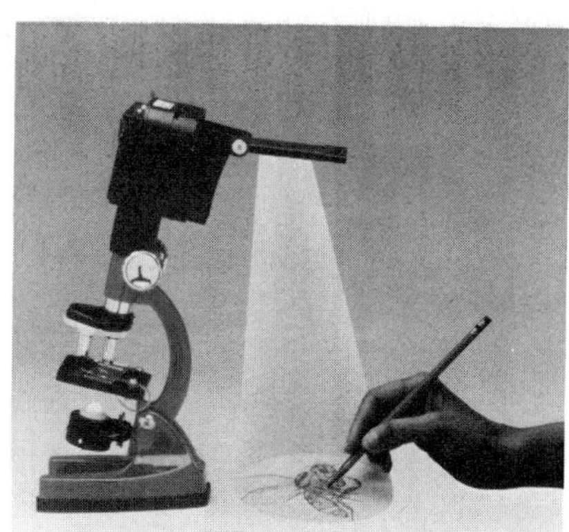

*Microscope*
*Tasco*

- solar print paper (International Playthings)
- fingerprint kit (Battat Experiments)
- periscope kit (Battat Experiments)

*Nature Print Paper*
*International Playthings*

## Games

During these years, game-playing comes into its own. Children are ready now for games that call for some strategy and competition. And playing by the rules is a lot easier, once the rules are agreed on. Before playing, a good deal of time is often spent negotiating and rewriting the rules, and there may be attempts to readjust them once the game begins.

By now, most children are much better prepared to accept the concepts of winning and losing. For those who still find losing painful, playing in teams can reduce some of the personal sting.

Games are enjoyed with both friends and family. Here are some of the best:

- **Strategy**
- Parcheesi (Selchow and Righter)
- Pente (Parker Brothers)
- Chess and Checkers (Milton Bradley)
- Othello (Milton Bradley)
- Stratego (Milton Bradley)

- **Word Games**
- Upwords (Milton Bradley)
- Boggle (Parker Brothers)
- Scrabble (Selchow and Righter)

- **Family Fun**
- Monopoly (Parker Brothers)
- Clue (Parker Brothers)
- Bingo (Milton Bradley)

- Made for Trade (Aristoplay)
- Music Maestro (Aristoplay)
- Wildlife Adventure (Ravensburger)

- **Math**
- Yahtzee (Milton Bradley)
- Battleship (Milton Bradley)
- Triomenos (Pressman)
- Uno (Uno)

- **Action**
- Twister (Milton Bradley)
- Nerf Ping-Pong (Parker Brothers)
- Power Jet Hockey (Coleco)

The family computer also lends itself to a variety of games. Many are pure shoot-'em-ups that depend on dexterity and speed with a joystick. Others do much more. Among the best for this age group are:

- *Arcade Machine* (Broderbund)
- *Bank Street MusicWriter* (Mindscape)
- *Bank Street StoryBook* (Mindscape)
- *Bank Street Writer* (Broderbund)
- *Factory* (Sunburst)
- *Flight Simulator* (Sublogic)
- *Koala Pad Painter* (Koala)
- *Movie Maker* (IPS)
- *Pinball Construction* (Electronic Arts)
- *Print Shop* (Broderbund)
- *Rocky's Boots* (Learning Company)

## *Puzzles and Manipulatives*

By now children are able to handle complex puzzles with small pieces. They don't need a frame, but they may need a little help now and then. Puzzles with one hundred to five hundred pieces are what enthusiastic puzzle fans will want. Such elaborate puzzles, of course, will require a surface that can be left undisturbed.

Children also are ready for some of the more complex cut-out model books. Though they may need some assistance, they enjoy the challenge of following directions and crafting toys, buildings, and vehicles that are worthy of display.

Consider also the ancient art of origami (Japanese paper folding) or moncuri (Japanese paper cutting). You'll find these in museum gift shops and many bookstores.

## *SUMMING UP*

These middle years are an important time for children to explore multiple interests and emerging talents. Through play, children continue to learn about themselves and others as social beings. With their increasingly refined dexterity, they hone their abilities as tool users and doers. Through active play they find the refreshment of using their whole bodies. They learn about the value of team play and cooperation as well as competition. In their pretend-play they find satisfaction in stepping outside themselves and into roles that would be otherwise impossible in reality. Not surprisingly, their toys need to be rich with the raw material for multiple modes of playing and learning.

# Conclusion

Given the glut of toys and the multitude of pressures to "Buy me! Buy me!," making choices in toyland is not as simple as it used to be. For parents it's more important than ever to become selective and knowledgeable consumers.

As parents we need to separate the business of selling toys from the important business of childhood, which is play. The toy business has grown radically by changing the way toys are designed and marketed. While there appear to be more choices, a closer look reveals that the toys available are more alike than not.

As a result, the playlives of children are being dominated by toys that fit marketing profiles rather than children's needs for multiple modes of playing. Many of today's shrink-wrapped toys also shrink the nature and quality of play.

While activists have focused our attention on the violent themes of children's TV and its related toys, the issues that parents face in toyland are broader than the debate over war toys.

With new technology comes sound- and light-activated toys—from mobiles in the nursery to action figures set in motion by the TV. Yes, they are innovative, but aren't they also intrusive?

In the business of selling not just to kids but also to parents and grandparents, the toymakers are merchandising feelings, fitness, imagination, and learning. Yet much of what they're selling is only in the press kits, not in the toys.

Borrowing from the language of academia, the toymakers have redefined the meaning of "open-ended playthings" to mean "you're never finished buying." So-called educational toys have reduced education to a series of right and wrong answers. Interactive play no longer means playing with others. In today's world the toy has become the player—the child has become the watcher. These days, imagination is not something children bring to their toys; it's what the toymakers provide with story, props, and full personality profiles. Yes, many are imaginative, but who's imagination is at play, the child's or the toymaker's?

For many adults, toys and play are just a frivolous part of childhood—nothing to be taken seriously. Others have heard or read that toys and play are important. So they assume that any kind of toys will do. Still others, who have heard that play is educational, assume that only certain kinds of toys can be taken seriously. Yet, in looking at toys and play across the years of childhood, parents will find that the value of toys and play changes like the clothes children wear. At every stage they need a rich variety of playthings to stimulate their minds and bodies.

We have also seen that much as we'd like to blame TV and the toymakers, parents themselves have played a part in fueling the "buy-me's!" Toys are not just playthings.

They've become tokens of guilt, marital conflict, affluence, aspirations, instant gratification. There's a lot of extra baggage packed into those shopping carts full of toys.

Television is not going away and neither are the pressures of the "buy-me" syndrome. This means that parents need to have a better understanding of the meaningful values of play in the lives of their children. To bypass some of the confusion, this book gives you a series of roadmaps to refer to as you travel through toyland. These roadmaps should give you a clear picture of the kinds of toys that are most apt to mesh with your child's expanding abilities and interests at each developmental stage. Of course, no one knows your child as well as you do. Trust that intimate knowledge. It too will help you predict the kinds of toys that fit your child's personal style, interests, abilities, and needs.

Remember, too, that even with the best roadmaps, there are bound to be detours and potholes along the way. So there are times that we say yes to a toy when we'd rather say no. And there are other times when we say no when it would be easier to say yes.

By understanding that play is not one thing but many, parents can provide a variety of props that empower children to become the real live-action figures in their own playlives.

# THREE

# *The Directory*

## *CATALOGUES*

For busy parents and grandparents, mail-order toy-shopping solves many problems. Not only does it save time, it also insulates you from the temptations of impulse buying. A few choice catalogues also provide you with access to toys you won't find in most toystores. It's like visiting a fine boutique with a small but select inventory. Several of the catalogues are from museum gift shops with heavy emphasis on hard-to-find science toys. Sharing the catalogue with kids may also open their eyes to other possibilities beyond the pop toys they're bombarded with on the TV screen.

In some cases the catalogue items plus shipping may cost more, but for many parents it's worth the price. Here are some of the catalogues that may be of interest. Catalogues are free unless otherwise noted.

Animal Town Game Co.
P.O. Box 2002
Santa Barbara, CA 93120

Many fine games not usually found in most toystores.

Chad's Rainbow
1778 N. Piano Road
Suite 120
Richardson, TX 75081
(214) 680-9787

Many of the toys from Brio, Ambi, and Kiddicraft mentioned in the book are in this catalogue. Toys from infants to preteens.

Childcraft Education Corporation
20 Kilmer Road
P.O. Box 3143
Edison, NJ 08818
(800) 631-5657

Well-selected playthings for children from infancy to schoolage. Ask for both home and school catalogues.

Community Playthings
Route 213
Rifton, NY 12471
(914) 658-3141

Sturdy wooden furniture, ride-ons, climbers, and blocks. Equipment is built for rugged use in schools and homes.

Discovery Corner
Lawrence Hall of Science
University of California
Berkeley, CA 94720

If you're looking for science toys, send for this one.

Discovery Toys
400 Ellinwood Way
Suite 300
P.O. Box 232008
Pleasant Hill, CA 94523-6008

An interesting collection of toys sold by home demonstrations for small groups.

Enchanted Doll House
Department BSC87
Manchester Center, VT 05255
(800) 362-3030

If you're looking for a dollhouse or furnishings, this catalogue has miniatures for both children and adult collectors. Send $3.00.

Heads Up
Early Learning Institute
445 East Charleston Road, Suite 9
Palo Alto, CA 94306
(415) 425-1155

Carefully selected toys for infants to age six. Each product comes with a skill card that gives suggestions to parents.

Metropolitan Museum of Art
Special Service Office
Middle Village, NY 11381
(516) 794-6270

Ask for their Presents for Children catalogue. Fine toys, games, books, and records. Discount for members.

Museum of Fine Arts
Catalogue Sales Department
P.O. Box 1044
Boston, MA 02120
(617) 427-7791

Small but select collection of books and art-related toys.

Museum of Modern Art
Mail Order Department
11 West 53rd Street
New York, NY 10019
(800) 346-0021

A few pages of this handsome catalogue feature playthings for children. Send $3.00; discount for members.

Nature Company
P.O. Box 2310
Berkeley, CA 94702
(800) 227-1114

As the name suggests, you'll find science toys and books for schoolage children. Gifts for teens and adults, too.

Pied Piper
The Gifted Children's Catalog
2922 North 35th Avenue, Suite 4
Drawer 11408
Phoenix, AZ 85061-1408

Plenty of unusual science toys for older kids as well as interesting playthings for the younger set.

Play Fair
1690 28th Street
Boulder, CO 80302

Describes its collection as nonviolent, nonsexist, and nonracist—for infants through schoolage children.

Smithsonian Institution
P.O. Box 2456
Washington, DC 20013
(703) 455-1700

Fascinating models and science toys, especially for schoolage children.

Toys to Grow On
P.O. Box 17
Long Beach, CA 90801
(213) 603-8890

An excellent catalogue with toys for infancy through school years. This is a division of Lakeshore, a major school supplier. You may want to request the school catalogue, too.

Several toy companies are also marketing their toys through the mail.

Lego Systems, Inc.
P.O. Box 640
Enfield, CT 06082
(203) 749-2291

Everything in the Lego line, from Dupo for toddlers to motorized sets for older kids. Great storage containers not generally found in toystores, and spare parts.

Learning Materials Workshop
58 Henry Street
Burlington, VT 05401
(802) 862-8399

Unique toys developed by a child development specialist.

Johnson and Johnson Child Development Toys
Stratmar Station
P.O. Box 7407
Bridgeport, CT 06650
(800) 433-3940 (in Connecticut)
(800) 334-0214 (elsewhere)

## *MANUFACTURERS AND DISTRIBUTORS*

The author wishes to thank the following toy companies for supplying photos for this book. The addresses are included so that readers may write directly if they cannot find a particular toy.

Ambi
Davis-Grabowski, Inc.
6350 N.E. 4th Ave.
P.O. Box 381594
Miami, FL 33138

Anatex Enterprises
14666 Titus Avenue, #7
Panorama City, CA 91402

Axlon
252 Humboldt Court
Sunnyvale
CA 94089

Ban Dai America, Inc.
4 Pearl Court
Allendale
NJ 07401

Badger Basket Company
616 North Court
Suite 150
Palatine, IL 60067

Binney and Smith, Inc.
1100 Church Lane
P.O. Box 431
Easton, PA 18044

Brio Scanditoy Corporation
6555 W. Mill Road
Milwaukee, WI 53218

Chasley
P.O. Box 19202
Seattle, WA 98109

Chicco
Artsana of America, Inc.
200 Fifth Avenue
Room 910
New York, NY 10010

Childcraft Education
Corporation
20 Kilmer Road
Edison, NJ 08818

Coleco Industries, Inc.
999 Quaker Lane South
West Hartford, CT 06110

Colorforms
Ramsey, NJ 07446

Creative Playthings, Ltd.
33 Loring Drive
Framingham, MA 01701

Creativity for Kids
Creative Art Activities, Inc.
2024 Lee Rd.
Cleveland Heights, OH 44118

Dakin and Company
P.O. Box 7746
San Francisco, CA 94120

Dolls by Pauline
International Design Toys
55 Pacella Park Drive
Randolph, MA 02368

Dynatoy International, Inc.
3594 West 1820 South
Salt Lake City
UT 84104

Effanbee Doll Corporation
200 Fifth Avenue
New York, NY 10010

Fisher-Price
Division of Quaker Oats
Company
636 Girard Ave.
East Aurora, NY 14052

Galt and Company, Inc.
63 North Plains Highway
Wallingford, CT 06492

Go Fly a Kite, Inc.
Route 151, Box AA
East Haddam, CT 06423

Gund, Inc.
1 Runyon Lane
Edison, NJ 08818

Hasbro, Inc.
1027 Newport Avenue
Pawtucket, RI 02861

Hedstrom Corporation
P.O. Box 432
Bedford, PA 15522

Inova Corporation
520 Madison Avenue
New York, NY 10022

International Playthings,
Inc.
116 Washington Street
Bloomfield, NJ 07003
(for Kiddicraft, Lardy,
Bambola, and others)

Johnson and Johnson Child
Development Toys
Grandview Road
Skillman, NJ 08558

Lauri, Inc.
Phillips-Avon, ME 04966

Learning Materials Workshop,
Inc.
58 Henry Street
Burlington, VT 05401

Lego Systems, Inc.
555 Taylor Road
Enfield, CT 06082

Little Tikes Company
2180 Barlow Road
Hudson, OH 44236

Lundby of Sweden USA, Inc.
14120 I and J Sullyfield Circle,
Chantilly, VA 22021

Malte Hanning Plastic
Ringvejen 51-53
DK-7900 Nykøbing M
Denmark

Manhattan Toy Company, Ltd.
33 West 17th Street
New York, NY 10011

Matchbox Toys
141 West Commercial Avenue
Moonachie, NJ 07074

Mattel Toys
5150 Rosecrans Ave.
Hawthorne, CA 90250

Milton Bradley
1027 Newport Ave.
Pawtucket, RI 02861

Ohio Art Company
P.O. Box 111
Bryan, OH 43506

Panosh Place, Inc.
200 Century Parkway
Mt. Laurel, NJ 08054

Play-Jour, Inc.
Suite 1024
200 Fifth Ave.
New York, NY 10010

Playskool, Inc.
1027 Newport Avenue
Pawtucket, RI 02861

Poppets
1800 E. Olive Way
Seattle, WA 98102

Ritvik Toy Corporation
Trimex Industrial Buildings
Route 11
Mooers, NY 12958

Woodkrafter Kits
Royal River Prints, Inc.
P.O. Box 808
Yarmouth, ME 04096

Skilcraft
A Division of Monogram Models, Inc.
8601 Waukegan Road
Morton Grove, IL 60053

Tasco Sales
7600 N.W. 26th Street
Miami, FL 33122

TC Timber
P.O. Box 42
Skaneateles, NY 13152

Tonka Toys
6000 Clearwater Drive
Minnetonka, MN 55343

Today's Kids
Highway 10 East
Booneville, AR 72927

Tyco Industries
200 Fifth Avenue
New York, NY 10010

View-Master Ideal Group, Inc.
P.O. Box 490
Portland, OR 97207

Wonderline, Inc.
200 Fifth Ave.
New York, NY 10010

Woodstock Chimes
Woodstock Percussion, Inc.
Route 1, Box 381A
West Hurley, NY 12491

Wright International, Inc.
Create It Educational Construction Sets
7600 Sixteenth Street, N.W.
Washington, D.C. 20012

## *PARENT ACTION GROUPS*

Here are the names of several parent action groups that you might want to join.

Action for Children's Television (ACT)
Auston Street
Newtonville, MA 02160

National Coalition on Television Violence (NCTV)
P.O. Box 2157
Champaign, IL 61820

Consumer Committee of the Americans for Democratic Action
3005 Audubon Terrace, NW
Washington, DC 20008

USA Toy Library Association
Suite 201
104 Wilmot Road
Deerfield, IL 60015-5195

Many communities have toy libraries where children can borrow toys. To find out how to organize, or to find the one nearest to you, write to the above address.

# *Index*


## A

accidents, 50–53, 67, 266
Action for Children's Television, 9, 10, 39, 296
activity boards, 73–74, 102
Activity Center, 74
Activity Gym, 196–97
Adica Pongo, 123, 184
advertising:
  children's perception of, 54, 162–63, 262–63
  expenditures on, 13–14
  television shows as, 7–9
  *see also* marketing
"affluenza," 44
AG Bear, 22, 211, 212
age labels, 29–31
All Star Sports Center, 224, 226
Ambi, 76, 104, 244, 297
American Academy of Pediatrics Task Force on Children and Television, 8
American Greeting Card Company, 7
Anatex Enterprises, 132, 297
Animal Grabber Puppet, 68
Animal Town Game Co., 294
Apokolips, 31
Aprica strollers, 170
Arcobeleno, 248–49
art materials, 107–9, 121–25, 139, 180–89, 244–45, 277
Axlon, 9, 22, 211, 212, 297

## B

Baby Mirror, 66
Baby's First Blocks, 74
Badger Basket Company, 169, 297
badminton, 274
balloons, 51
balls, 76–77, 90, 117–18, 197–98, 223–26
Balls in a Bowl, 77, 81
Ban Dai America, Inc., 279, 281, 297
*Bank Street Music-Writer,* 271
Bank Street Survey, 53, 159
Barbie, 16, 129, 170, 261, 267, 268, 270

baseball cards, 20
Bath Baby, 129
Bath Time Water Works, 120–21
beach balls, 90, 118
Bearbaby, 90
Bettleheim, Bruno, 148–49
bicycles, 193, 219–20, 274
Big Bird Story Magic, 211
Big Sandbox, 199, 201
Binney and Smith, Inc., 107, 108, 181, 182, 245, 277–78, 297
binoculars, 229, 230, 256
bird-watching, 229–31
blocks, 17, 28, 46, 74–75, 97–100, 135–36, 174–79
Block Sensations, 74, 75
Blue Bird, 100
board games, 206–7
Bonecrusher, 31
books, 103, 136–37, 209–12, 252–54
boxes, 118, 129
brand names, 52
bribery, 42–43
Brio-Mec, 242
Brio Scanditoy Corporation, 167, 168, 208, 238, 243, 268, 294, 298
Bristle Blocks, 136
Bruna Fabric Blocks, 75
brushes, 184
Bubble Mower, 127, 193, 194
Builda Helta Skelta, 248–49
Bunny Builders, 135, 136
Busy House, 95
Busy Shapes, 133, 134
Butterball Baby, 129
Buzz Saw Hordak, 33

## C

Cabbage Patch Kids, 20–21, 46, 167, 169, 237, 269
cameras, 256
Candyland, 206
Cannonball Express, 272–73
Capsela, 279, 280
Captain Power, 9
Care Bears, 9, 15, 21
Cargo Carrier, 173
Carles, Eric, 136
Carlson Capitol Manufacturing Company, 138
cars, 76, 88, 173–74, 272–73
Casey the Robot, 210–11, 214
catalogues, 293–96
CBS Toys, 14
Chad's Rainbow, 294
chalkboards, 245
Charrin, Peggy, 9
Chasley, 271, 298
checkers, 251
Cherubin, 238
Chicco, 298
Childcraft Education Corporation, 91, 118, 121, 164, 167, 185, 187, 188, 197, 198, 199, 203, 208, 247, 248, 294, 298
childproofing, 84–85
children:
  desensitization of, 37
  learning of values by, 55–57
  perception of advertising by, 54, 162–63, 262–63
  *see also* early school years; infants; middle years; preschoolers; toddlers
*Children's Business,* 11
Children's Television Workshop (CTW), 47
Chimalong, 190, 252, 271
Chime Ball, 76
Chime Bird, 76–77
Chinese checkers, 251
*Choosing Books for Kids,* 137
Chugga Chugga Choo Choo, 116
Chutes and Ladders, 251
clay, 186–87, 244
climbing, 90–92, 115
Clippety Clop, 198
Clown Rattle, 68
Cog Labyrinth, 208
Coleco Industries, Inc., 21, 34, 169, 298
collages, 187–89
collecting, 18–20, 261–63
Colorforms, 187, 298
coloring books, 17
Color Peg Board, 208
Combination Sink and Stove, 166
Community Playthings, 294
Compact Kitchen, 166
computers, 213–16, 253–54, 271, 285
construction toys, 97–100, 135–36, 179–80, 241–43, 279–81
Construction Vehicle, 241, 242
Construx, 279–80
Consumer Committee of the Americans for Democratic Action, 296
consumerism, 53–55
Consumer Product Safety Commission, U.S. (CPSC), 50, 53, 67
Cookie Counter, 253
cooking, 255–56, 282
Cootie, 206
Corniche convertibles, 23

Corning Ware, 143
Corn Popper, 89
Cotton Reels, 206
Counter Balance Support Company, 138
crafts, 246–47, 276–77
Crayola Crayon Case, 245
crayons, 17, 107–8, 182, 244
Crazy Combo, 190
Create It, 242, 299
Creative Lettering Set, 277–78
Creative Playthings, Ltd., 14, 91, 196, 298
Creativity for Kids, 233, 298
crib gyms, 68–69
crib toys, 64
Curiosity Ball, 77
Curious George, 210

## D

Dakin and Company, 67, 76, 172, 269, 298
Dancing Animal Music Box Mobile, 64
Dapper Dan, 134
dashboards, 130
daycare, 141
D. Compose, 31, 32
Delta Force, 34
desensitization, 37
Designer Kit for Vehicles, 278
Dietz, William H., 8–9
dinosaurs, 170
Discovery Corner, 279, 294
Discovery Toys, 294
dishes, 127, 143, 165
diving masks, 228
Dr. Hacker, 31
Dr. Terror, 32
Doll Buggy, 105
doll carriages, 105, 129, 169, 193
dollhouses, 166, 239–40, 267–68, 294
dolls, 29–30, 92–94, 95, 105, 129, 134, 167–69, 236–38, 267–69
Dolls by Pauline, 167, 169, 298
dressing, 134
dress-up clothes, 130, 164
drop-and-pick-up game, 79
Duplo blocks, 99–100, 136
Dynatec International, Inc., 28, 298

## E

Early Learning Institute, 294
early school years, 25–26, 217–58
  play of, 217–19, 232–34
  real experiences and, 254–57
  social development of, 218, 234–36
  toys for, 219–57
  TV and, 234–36
easels, 182–84
educational toys, 12–13, 24–26, 218
  *see also* intellectual development
Effanbee Doll Corporation, 129, 298
electronic plush toys, 14–15, 210–12
Electronic Supersound Driver, 164–65
electronic toys, 26, 214–16, 252–53
Elmers glue, 182
Enchanted Doll House, 294
Engelhardt, Tom, 11
Erector Sets, 279
Etch-a-Sketch, 188
ethnic dolls, 169
Evil-Lyn, 33
exercise mats, 225, 226, 274
expectations, 40, 155
extensions, 6, 15–16

## F

fairy tales, 30, 37, 148–50
Fantastix, 279, 281
farms, 139
Fashion Plates, 247
fast-food chains, 51
Federal Communication Commission (FCC), 9
finger paint, 121–22, 185–86
Fisher-Price, 64, 68, 73, 74, 76, 89, 93, 94, 100, 127, 128, 131, 164, 165, 166, 190, 193, 194, 195, 222, 223, 253, 256, 272, 276, 277, 279, 298
fitness equipment, 226–27
fitness programs, 23
Flip Flop Tool Box, 98
floor toys, 134
Flower Rattle, 76
Fluppy Puppies, 21–22
Fold 'n Go Activity Quilt, 72
Follow Through Programs, 234
Frisbees, 227
Fun-To-Learn, 215
furniture, 106–7, 127, 202–3

## G

Galt and Company, Inc., 87, 95, 126, 127, 175, 184, 204, 205, 208, 246, 247, 276, 277, 298
games, 206–7, 223–26, 250–52, 260–61, 274–75, 284–85
gates, 90
gender-based toys, 171–72, 237–38, 266–70
General Mills, 14
gifts, 40–44
  unwanted, 49, 70
GI Joe, 16, 31, 34, 38, 237, 261, 270
Gilkeson, Elizabeth, 234–35
Glug-Glug boat, 198–99
Gobots, 9, 10, 15, 18, 31
Go Fly a Kite, Inc., 275, 298
Goldberger Dolls, 129
Golden Ribbon Playthings, 169
Grandma's Trunk, 164
Grandpaw Bear, 211
Gross Out Gang, 34
gross toys, 34–36
guilt, 41–42
Gummie Bears, 36
Gund, Inc., 67, 68, 70, 71, 298
Gymboree-style classes, 23
gyms, 90–91, 195–97, 227
  crib, 68–69

## H

hammer boards, 98
handmade toys, 70
Hasbro, Inc., 7, 9, 13–14, 16, 32, 167, 172, 256, 260, 274, 298
Headstrom Corporation, 147, 197, 198, 298
Hello Kitty, 14
He-Man and Masters of the Universe, 7, 9, 10, 11, 19, 31, 33, 131, 159, 172
hinged toys, 95–96
Hocus Pocus Set, 271
holograms, 10
Hoppety-Hop Ball, 197, 198
Hot Wheels, 9, 88, 173
Hot Wheels Car Wash, 239
household items, 48, 84–85, 96, 102, 118, 129, 177, 187, 199
housekeeping toys, 127, 165–66
Hug-a-Planet, 198
Huggy Bean, 169

## I

ImaginAction, 74
imaginary playmates, 152
Imaginetics Magnetic Blocks, 179, 180
infants, 26–27, 61–82
  fitness programs for, 23
  independence of, 78–79
  overstimulation of, 62
  play of, 78–80
  toys for, birth to 6 months, 63–73
  toys for, 6–12 months, 73–81
Inhumanoids, 32
Integrity Design, 279, 281
intellectual development, 25–26, 144–45, 152–54, 155–56, 176–77
interactive toys, 27
International Playthings, Inc., 75, 120, 135, 200, 207, 225, 284, 298

## J

jack-in-the-boxes, 95
Jem, 18
Jeux et Jouets Educatifs, 299
jigsaw puzzles, 248
Johnson and Johnson Child Development Toys, 26–27, 66, 68, 73, 76, 77, 78, 81, 89, 94, 120, 296, 298
juggling, 271–72
jump ropes, 227

## K

kaleidoscopes, 208
Kenner, 14, 21, 32, 186, 247
Key House, 133, 134
keys, 105, 133–34
Kiddicraft, 67, 75, 81, 100, 101, 248, 249, 294
Kiddi Links, 95
Kids Puzzle, 249, 250
Killgallon, Larry, 14
Kindercolor Express, 132
Kindergund, 67, 68, 70
kites, 275
Ko Ko, 269
Kuhn, Jeep, 10–11

## L

Lace Shapes, 206–7
language, 97, 112–14, 151–52

lap desks, 188
lap toys, 66–68
laser guns, 35
Lauri, Inc., 240, 249, 250, 298
Lazer Tag, 35
Learning Curves, 77
Learning Materials Workshop, Inc., 248–49, 296, 298
Lego Systems, Inc., 46, 99–100, 136, 179–80, 241, 279, 296
Let's Cook, 126
licensed toys, 5–7, 46–48, 131–32, 137, 142–43, 192, 210
Lincoln Logs, 21, 241–43
Linky Dinks, 67
Lipscomb, Eloise, 44
Little Looker Pocket Microscope, 228–29
Little People Environments, 131
Little Professor, 253
Little Tikes Company, 88, 92, 98, 105, 106, 115, 116, 119, 130, 131, 165, 166, 183, 184, 190, 196, 199, 201, 298
Lottino, 207
Lotto, 206
Lundby of Sweden USA, Inc., 240, 267, 268, 298

## M

Mad Scientist Monster Lab, 34
Magic Brushes, 184
Magic Slate, 188
magic tricks, 271
Magic Vac, 127, 128
Magna-Doodle, 188
Magnctic Picturc Blocks, 187
Magnetic Train, 168
magnets, 208
magnifying glasses, 208
makeup, 271–72
Malte Hanning Plastic, 299
Manhattan Toy Company, Ltd., 170, 299
manipulative toys, 73–76, 94–109, 132–36, 207–8, 286
manufacturers and distributors, 297–99
markers, 108, 185, 244
marketing, 5–16
  "experts" and, 12
  as limiting choices, 10–11
  long-term impact of, 39
  promotion and, 11–13
  TV-toy connection in, 8–10, 15
  of violent toys, 36
  *see also* advertising; licensed toys
Masters of the Universe, *see* He-Man and Masters of the Universe
Match a Balloon, 206
Matchbox Toys, 12, 88, 173, 239, 299
Mattel Toys, 9, 11, 14, 16, 19–20, 31, 32, 117, 170, 173, 239, 299
media tours, 12
Medical Kit, 164
Mega Blocks, 136, 178, 180
Mega Tech Blocks, 153
Melody Push Chime, 88–89
Memory, 206
Messy Play and Hobby Tray, 121–22, 185
Metropolitan Museum of Art, 295
microscopes, 228–29, 282, 283
middle years, 259–86
  play in, 259–70
  social pressure and, 263–65
  toys for, 271–86
Mighty Tonka, 173
Mighty Toys Explorama, 11
Milton Bradley, 14, 253, 299
mirrors, 66, 74, 75
mobiles, 26–27, 64–65
moncuri, 286
Mother Duck, 81
Mother Goose, 211, 213
movies, 34
Mucous Pukous, 34
Mumm-Ra, 32
M.U.S.C.L.E.s, 19–20
Museum of Fine Arts, 295
Museum of Modern Art, 295
music, 103–4, 119, 189–90, 271
music boxes, 72, 190
My Buddy, 167, 169
My First Car, 76
My First Phone, 104
My Little Pony, 19, 236–37, 270
My Shape 'n Stir Pot, 96, 97

## N

National Coalition on Television Violence (NCTV), 39–40, 160, 296
Nature Company, 295
Nature Print Paper, 284
needlepoint, 246, 276
Nerf balls, 197

Nerf Ping-Pong Set, 274
*Newsweek,* 21
*New York Times,* 8
Nickel Plate Road Set, 272
Noah's Ark, 166, 167
Nose Ark, 34

## O

Ohio Art Company, 14, 188, 274, 275, 299
Ophelie, 268
origami, 286
overstimulation, 62

## P

paints, 121–23, 182–86, 244, 277–78
Panosh Place, Inc., 299
paper dolls, 246
papier-mâché, 244, 272
Paraphernalia For Pretending, 233
Parcheesi, 251
parents, 40–46, 49
  action groups, 296
  alternate toys encouraged by, 45–46, 48
  bribery and, 42–43
  child in, 43–44
  guilt of, 41–42
  infants' play and, 78–80
  limit-setting of, 44–45, 266
  middle-years play and, 260, 265, 266
  older-toddlers' play and, 138–39, 144–45
  preschoolers' play and, 157–58
  unwanted gifts and, 49
  values and, 55–57
  violent toys and, 158–62
  young toddlers' play and, 84–85
Parker Brothers, 197, 274
Party Kitchen, 166
Pattern Printing set, 247
*Pat the Bunny,* 103
Paulsen, Terri, 11
peekaboo, 80
Peg Pan, 207
penpals, 262
pens, 108
Peter Rabbit, 210
pets, 231
physical development, 22–24, 114–25
Picture Dominos, 206
Pied Piper, 295
Piglets, 68
Piky, 248
Ping-Pong balls, 77, 90
Pipeworks, 197, 243
plants, 229
Plasticine, 124, 187
play, xiv–xxi, 16–31
  of early school years, 217–19, 232–34
  emotional content of, 21–22
  imagination in, 27–29
  of infants, 78–80
  learning and, 24–26
  in middle years, 259–70
  of older toddlers, 114–15, 125–32, 138–39
  old vs. new, 263–64
  of preschoolers, 146–58, 191–92
  pretend, xx, 104–7, 125–32, 147–48, 164–70, 232–54
  research and, 26–27, 62
  transformation in, 17–18, 22
  as "work," 24–26
  of young toddlers, 84–85, 109–10
Play Boxes, 203
Play Buckets, 100, 101
Play-Doh, 109, 186
play dough, 109, 123–24, 186
Play Fair, 295
playgroups, 142
playhouses, 166
Play-Jour, Inc., 279, 280, 299
playpens, 80–81
Play People House, 166
Playskool, Inc., 14, 25, 70, 72, 76, 77, 81, 95, 96, 133, 134, 136, 164, 169, 197, 198, 200, 210, 214, 215, 241, 243, 256, 299
Play Table, 188
plush toys, *see* electronic plush toys; soft toys; stuffed animals
Pogo Bal, 260, 274
Poppets, 172–73, 241, 299
Popples, 18
Pound Puppies, 269
Power Shirt, 28
Preschool Building Set, 153
preschoolers, 35–36, 146–216
  computers and, 213–16
  intellectual development of, 25–26, 152–54, 155–56, 176–77
  language of, 151–52
  play of, 146–58, 191–92
  social development of, 150–51, 156–57, 176
  toys for, 163–216

pretend play, xx, 104–7, 125–32, 147–48, 232–54
props for, 164–70
printing sets, 247
Prismatics, 248
Puffalumps, 93, 94
pull toys, 89, 117, 193
punching bags, 197
puppets, 172–73, 240–41, 272
push toys, 88–89, 105, 127, 129, 193
puzzles, 132–36, 139, 204–6, 248–50, 286

## R

racing sets, 239
Radar Ball, 275
Radio Flyer, 117
Radio Flyer Scooter, 221
Raggedy Ann, 167
Rainbow Brite, 11, 36, 54
Rambo, 34
rattles, 66, 68, 73, 76
reading, 25–26, 78
Real Baby, 238
Red Rings Rattle, 73, 76
Remote Control Racer, 273
research, 26–27, 62, 160, 235
Rhythm Roller, 76, 78
riding toys, 23, 51, 87–88, 116–17, 192–95, 219–21
ring-and-post toys, 132
Rings and Rollers, 94
Ritvik Toy Corporation, 136, 153, 178, 299
Robotech, 12
Rocket Ring Toss, 225
rocking chairs, 202
rocking horses, 88, 197
Rollercoaster, 132
Rubber Stamp Printing Set, 247

## S

safety, 49–53, 64, 68–69, 84, 86, 90, 92, 137–38, 180, 220, 266, 274
sandboxes, 199
Sand Castle Play Set, 200
Sand Storm, 33
schools, toys provided to, 12–13
science materials, 282–84, 294, 295
scissors, 181, 182
scooters, 220–21
Scotch Tape, 182
Scrabble People, 153
Sears Catalogue, 91
See-Through Buckets, 202
Sesame Street, 47–48, 172
Sesame Street Busy Poppin Pals, 134
sewing, 246
sewing cards, 206
shape sorters, 96–97, 133
Shape Sorter Transporter, 89
sharing, 140–41, 150–51
She-Ra, 131, 172
Shooting Stars, 271
Shuffletown, 28
siblings, 86
Simon, 253
Sinus Slimus, 34
skateboards, 266
skates, 193–94, 221–23
Skilcraft, 179, 186, 273, 275, 283
skis, 223
Small Hot Wheels, 173
Smithsonian Institution, 295
Smurfs, 143
Snappy Snail, 87
Snoopy, 134
snow gear, 223
Snow White, 21
So Big Crayons, 107, 108
social development, xx, 140–45, 150–51, 156–57, 176, 218, 234–36
Softina, 129
Soft Touch, 208
soft toys, 69–71, 129, 167, 267–69
*see also* stuffed animals
Sorting Blocks, 208
Sound Puzzle Box, 96
Spacewarp, 279, 281
Spirograph, 247
sports, 260–61, 273–75
Spot books, 136
squeeze toys, 67
Squish Paints, 123, 184
Stacking Sand-and-Water Wheels, 199
Stack 'n Store Nesting Cubes, 98
stairs, 90
Star Rings Rattle, 76
Stitch Bear, 70, 71
storage, 52, 137–38, 202–3
Strawberry Shortcake, 6–7, 15, 270
Stuffed animals, 92–94, 105, 130, 170–71, 236–38
*see also* soft toys
Super Building Blocks, 99
Supernaturals, 10
Sutton-Smith, Brian, 41
swings, 92, 115

Sylvanian Doll House, 239
symbols, 18, 154–55

## T

talking toys, 210–12
Talk 'n Play, 214–15
Tap-A-Tune Piano, 119, 190
tape recorders, 271
tapes, 209–12, 252–54
Tasco Sales, 228, 230, 283, 298
TC Timber, 166, 167, 207, 239, 299
*Teach Me Reader,* 25–26
Teach-Me Watch, 256
tea parties, 105–6, 165
Tech Force, 9
Teddy Ruxpin, 14–15, 210–11
teething rings, 67, 76
telephones, 104
telescopes, 282, 283
Television Bureau of Advertising, 13
television programs:
  age labels and, 30
  common background provided by, 234
  as fairy tales, 30, 37, 148–50
  impact on children of, 235–36
  licensed toys from, 5–7, 46–48, 131–32, 137, 142–43, 192, 210
  product-driven, 7–9, 15, 28, 30
  selective viewing of, 44–45
  violent, 31, 37–39, 158–62
tempera paint, 244
Texas Instruments, 253
theatrical shows, 271–72
throw-away toys, 20–21
Tike Treehouse, 115
Timex Teach-Me Watch, 256
Tinker Toys, 180
Today's Kids, 224, 226, 298
Toddler Gym, 91–92
toddlers, older, 111–45
  independence of, 111–12
  intellectual development of, 144–45
  language of, 112–14
  licensed toys and, 131–32, 137, 142–43
  physical development of, 114–25
  play of, 114–15, 125–32, 138–39
  social development of, 140–45
  toys for, 114–37
toddlers, young, 83–110
  childproofing for, 84–85
  language of, 97, 103
  play of, 84–85, 109–10
  siblings of, 86
  stairs and, 90
  toys for, 86–110
Toddler Swing, 92
Toddle Tots Family Car, 88
Toddle Tots Family House, 130, 131
Tonka Toys, 9, 11, 14, 173, 299
Toot 'n Toddle Taxi, 87–88
Tossed Salad Juggling Set, 271
Touch 'Em Clutch Ball, 77
*Toy and Hobby World,* 7, 14
toy chests, 137–38, 202–3
Toy Fair, 13, 15, 20–21, 143
Toy Manufacturers of America (TMA), 7, 38
toys:
  alternate forms of, 45–46, 49
  big-company domination of, 13–15
  glut of, 3–6
  homogeneity of, 4, 10–11
  as line extensions, 6, 15–16
  marketing of, 5–16
  as overstimulating, 30
  retail sales of, 8
  safety of, 49–53; *see also* safety
  short lives of, 20–21
  TV-activated, 9–10
  *see also specific toys and types of toys*
Toys-R-Us, 15
Toys To Grow On, 164, 166, 225, 295
trains, 167, 168, 272–73
transformations, 17–18, 22, 154–55
Transformers, 7, 10, 18, 29, 31, 174, 270
transportation toys, 173–74, 272–73
traveling toys, 72–73, 187–88
tricycles, 192
trucks, 173
Tub Pals, 81
Tunnel of Fun, 118
Twin Faces, 76
Twistrack, 239
Two Potato Clock, 283
Tyco Industries, 100, 153, 239, 241, 243, 273, 299

## U

USA Toy Library Association, 296

*Uses of Enchantment, The* (Bettleheim), 148

## V

values, 55–57
video cameras, 272
video cassettes, 213
View-Master Ideal Group, Inc., 188, 211, 299
violence, 5, 31–40, 158–62, 263–64
Visionaries, 10
volleyball, 275
Voltron, 31

## W

Wacky Bat and Ball, 198, 199
wagons, 117, 193, 220–21
walkers, 80–81, 87–88
Walker Wagon, 87
walkie-talkies, 256–57
*Wall Street Journal,* 44
watches, 256
watercolors, 244, 277
water toys, 35, 81–82, 100–102, 119–21, 129, 167, 198–99, 227–28
weaving looms, 276, 277
Wee Wheels, 76
Weiner, Richard, 11
Wicker Cradle, 169–70
Wind-Up Sea Plane, 198, 200
Wonderline, Inc., 87, 198, 299
Wooden Unit Blocks, 175
Woodkrafter Kits, 299
Woodstock Percussion, Inc., 252, 299
woodworking, 243–44
*Working Mother,* 20
working mothers, 41, 141
Worlds of Wonder, 14–15, 35, 75, 210, 211, 213
Wright International, Inc., 243, 299
Wuzzles, 15, 36

## Z

Zap-It Liquidator, 35
Zero Gravity Cliff Hangers, 273

## ABOUT THE AUTHOR

Joanne F. Oppenheim is associate editor for the Media Group at the Bank Street College of Education in New York and author of *Choosing Books for Kids* and *Raising a Confident Child*, among other works.